AF352933

Probability and Stochastic Processes with Applications to Communications, Systems and Networks

Probability and Stochastic Processes with Applications to Communications, Systems and Networks

Editors

Gurami Tsitsiashvili
Alexander Bochkov

MDPI • Basel • Beijing • Wuhan • Barcelona • Belgrade • Manchester • Tokyo • Cluj • Tianjin

Editors
Gurami Tsitsiashvili
Far Eastern Branch of Russian
Academy Sciences
Russia

Alexander Bochkov
Signaling and Telecommunications in Railway
Transportation – NIIAS, JSC
Russia

Editorial Office
MDPI
St. Alban-Anlage 66
4052 Basel, Switzerland

This is a reprint of articles from the Special Issue published online in the open access journal *Mathematics* (ISSN 2227-7390) (available at: https://www.mdpi.com/si/mathematics/Probab_Stoch_Process_Appl_Commun_Syst_Netw).

For citation purposes, cite each article independently as indicated on the article page online and as indicated below:

LastName, A.A.; LastName, B.B.; LastName, C.C. Article Title. *Journal Name* **Year**, *Volume Number*, Page Range.

ISBN 978-3-0365-6485-2 (Hbk)
ISBN 978-3-0365-6486-9 (PDF)

Contents

About the Editors

Gurami Tsitsiashvili

Gurami Tsitsiashvili, Dr. Science, is currently a Professor at the Institute for Applied Mathematics, the Far Eastern Branch of the Russian Academy of Sciences, Russia. He received an M. Sc. in Mathematics and Physics from Moscow Physic Technical Institute in 1972 and a Ph.D. in Mathematics and Physics from the Institute for Applied Mathematics, the Far Eastern Branch of the Russian Academy of Sciences in 1976. In 1992, he achieved Dr. Sci. in Mathematics and Physics with a Rehabilitation Thesis "Decomposition Analysis of Complex Systems". In 1972, he was stationed in the Far East division of the Russian Academy of Sciences, where he has worked until now, starting as a trainee-researcher and later becoming the Deputy Director of Science at the Institute of Applied Mathematics of the RAS. He is the author of more than 200 works, including four monographs, one textbook and more than 110 articles in international peer-reviewed magazines. He studies complex systems, mass service system stability, cooperativity in multiple stochastic systems and multiplicative theorems for queueing networks.

Alexander Bochkov

Alexander Bochkov, Dr. Science, is currently a Scientific Secretary at the Research and Design Institute for Information Technology, Signaling and Telecommunications in Railway Transportation—NIIAS, JSC, Co-founder of Gnedenko-Forum. Alexander Bochkov graduated from the faculty of the Moscow Aviation Institute (MAI) in 1994. After graduation, he worked as a researcher at the Scientific-Technological Complex System Analysis, Russian Research Center "Kurchatov Institute", Scientific Institutes and organizations of Gazprom. In 2000, he produced a dissertation for the degree of Ph.D. (candidate of technical sciences), and in 2019, he produced a dissertation for the degree of Dr. Science. Dr. Bochkov participated in many Russian and international conferences and scientific schools and has published in Russian and foreign journals more than 150 scientific papers and three monographs. His research interests are life-support systems, water purification electrochemical systems, non-stationary processes, risk assessment, analysis and management, system analysis and operation research and vulnerability and survivability of large-scale systems.

Editorial

Preface to the Special Issue on Probability and Stochastic Processes with Applications to Communications, Systems and Networks

Alexander Bochkov [1,*] and Gurami Tsitsiashvili [2]

[1] JSC NIIAS, 109029 Moscow, Russia
[2] Institute for Applied Mathematics, Far Eastern Branch of Russian Academy of Sciences, 690041 Vladivostok, Russia
* Correspondence: a.bochkov@gmail.com or a.bochkov@vniias.ru

Citation: Bochkov, A.; Tsitsiashvili, G. Preface to the Special Issue on Probability and Stochastic Processes with Applications to Communications, Systems and Networks. *Mathematics* **2022**, *10*, 4665. https://doi.org/10.3390/math10244665

Received: 5 December 2022
Accepted: 7 December 2022
Published: 9 December 2022

Publisher's Note: MDPI stays neutral with regard to jurisdictional claims in published maps and institutional affiliations.

This Special Issue is devoted to probability, statistics, stochastic processes, and their different applications in systems and networks analysis. The Special Issue will include works related to the analysis and applications of different queuing models, which begin with general approaches to modeling queuing systems and networks. Significant attention will be devoted to the analysis of probabilistic and statistical methods in telecommunication; asymptotic analysis of queuing networks in the condition of a large load will be considered since original approaches are being developed in the asymptotic analysis of queuing networks in the condition of a large load and in the calculation of distributions in retrial queuing systems. We welcome considerations of general complex networks and their structures in terms of, e.g., topology and graph theory; mathematical methods and models in smart cities; exclusive statistical methods, such as statistical estimates in bio/ecology, medicine, and neural networks; and works that estimate parameters in complex technical systems, etc.

The authors' geographical distribution is shown in Table 1; the 21 authors are from eight different countries. Note that it is usual for a paper to be written by more than one author and for authors to collaborate with authors with different or multiple affiliations.

Table 1. Geographic distribution of authors by country.

Country	Number of Authors
Canada	3
Russia	10
India	1
Czech Republic	2
Saudi Arabia	2
Egypt	1
Korea	1
UK	1

Mass serving systems are widely used in many areas of real life. While single-server queue systems work in some cases, multi-server systems can efficiently handle the most complex applications. Multi-server mass service systems (compared to well-designed single-server systems) are more complex and more difficult to handle, especially when the arrival time distribution is arbitrary. The paper [1] is devoted to the analytical and computational analysis of queue length distributions for a complex multi-server mass service system. Introducing a quorum further complicates the model. In view of this, a two-dimensional Markov chain must be employed. It now appears that this system has not been considered so far. An elegant closed-form analytical solution and an efficient algorithm for obtaining a queue length distribution in three different epochs are presented.

Specialists in medical and zoogeography, mining, applications of meteorology to field problems, etc., have considerable interest in large or extreme outliers in sets of empirical information. For the following purposes, specialists are important: the essential importance of large emissions, the fear of errors in the study of large emissions by standard and previously applied methods, the speed of information processing, and the ease of interpretation of the results obtained. To meet these requirements, algorithms for interval pattern recognition and accompanying auxiliary computational procedures were developed in [2]. These algorithms were developed for specific samples provided by users (short samples, the presence of rare events in them, or the difficulty of constructing interpretation scenarios). What they have in common is that original optimization procedures are constructed for them or known optimization procedures are used. The authors present a series of results on the processing of observations through the extraction of large outliers, both in the time series and in planar and spatial observations. The algorithms presented in [2] are fast and sufficiently valid in terms of specially selected indices and have been tested on specific measurements and accompanied by meaningful interpretations.

In [3], the authors present an alternative and simpler approach to finding stationary distributions of the number of jobs for a mass service model with finite space using roots of its own characteristic equation. The main advantage of this alternative process is that it provides a unified approach to working with both finite-buffer and infinite-buffer systems. The queue length distribution is obtained both at the departure epoch and at the random epoch.

Typically, a complex system consists of various components that are usually subject to service policies. In [4], the authors consider systems containing components that are under preventive maintenance and repair maintenance. Preventive maintenance is treated as a failure-based preventive maintenance model in which a complete update is implemented after every nth failure occurs. It proposes an imperfect corrective maintenance model in which each repair worsens the lifetime of a component or system, whose probability distribution gradually changes by increasing the failure rate. The paper demonstrates reliability mathematics for quantifying unavailability. A model of the renewal process involving preventive maintenance based on failure arises from a new corresponding renewal cycle, which is denoted as the real aging process. Imperfect corrective maintenance leads to an undesirable increase in the unavailability function, which can be corrected by a correctly chosen failure-based preventive maintenance policy, i.e., replacing the correctly chosen component considering both cost and unavailability after the n-th failure occurs. The number n is considered the decision variable, while cost is the target function in the optimization process. The paper describes a new method for finding the optimal preventive maintenance policy based on failures for a system considering a given reliability constraint. The decision variable n is optimally chosen for each component from a set of possible realistic maintenance policies. The authors focus on a discrete maintenance model in which each component is implemented in one or more maintenance modes. A fixed value of the decision variable determines one mode of service as well as the cost of the mode. The system optimization process requires computation time because if the system contains k components, each with three service modes, 3k service configurations need to be estimated. Discrete service optimization is shown for two systems taken from the literature.

Today's smart grids make it possible to efficiently manage energy supply and consumption while avoiding various safety risks. System disturbances can be caused by both natural and man-made events. Operators must be aware of the different types and causes of power system disturbances to make informed decisions and respond appropriately. Research [5] proposes a solution to this problem with a deep learning-based attack-detection model for power systems that can be trained using data and logs collected from vector measurement units (PMUs). Creating properties or specifications is used to create features, and the data are sent to various machine learning methods, of which the random forest was chosen as the main classifier by AdaBoost. Data from simulated energy systems from open sources are used to test a model containing 37 case studies of energy system events. The proposed model was compared to other layouts on various evaluation metrics. Simulation results showed that this model provides a detection rate of 93.6% and an accuracy rate of 93.91%, which is higher than existing methods.

In [6], a variant of group testing (GT) models, called noise threshold group testing (NTGT), is considered, in which if there is more than one defective sample in the pool, its test result is positive. The authors are dealing with a variant model of GT in which, as in the diagnosis of COVID-19 infection, not only do false positives and false negatives occur if the virus concentration falls below the threshold, but unexpected measurement noise can change the correct result above the threshold to become incorrect. The authors aim to determine how many tests are needed to recover a small set of defective samples in such an NTGT problem. To do this, they find necessary and sufficient conditions for the number of tests needed to recover all defective samples. First, Fano's Inequality was used to obtain a lower bound on the number of tests needed to satisfy the necessary condition. Second, an upper bound was found using the MAP decoding method, which leads to a sufficient condition to recover defective samples in the NTGT problem. As a result, the authors show that the necessary and sufficient conditions for successful reconstruction of defective samples in the NTGT coincide. In addition, they show a tradeoff between the percentage of defective samples and the density of the group matrix, which is then used to construct the optimal NTGT structure.

The paper [7] introduces a stochastic process of an inhomogeneous Markov system in a stochastic environment in continuous time (S-NHMSC). The ordinary inhomogeneous Markov process is a special case of S-NHMSC. The author studied the expected population structure of the S-NHMSC, the first central classical problem of finding the conditions under which the asymptotic behavior of the expected population structure exists, and the second central problem of finding which expected relative population structures are possible limits if the limiting vector of input probabilities into the population is controlled. Finally, the rate of convergence is studied.

In various areas of human activity, there is inevitably a need to select the best (rational) courses of action from the alternatives proposed. In the case of retrospective statistics, risk analysis is a convenient tool for solving the choice problem. However, when planning the growth and development of complex systems, a new approach to decision making is needed. The article [8] deals with the concept of risk synthesis in comparing alternatives for the development of a special class of complex systems, which the authors call smart expansive systems. "Smart" in this case implies a system capable of balancing its growth and development, considering possible external and internal risks and constraints. Smart expansive systems are considered in the quasi-linear approximation and under stationary problem-solving conditions. In the general case, when the alternative comparison is not the object itself, but some scalar way of determining risks, the problem of selecting the objects most exposed to risk is reduced to the evaluation of weights of factors influencing the integral risk. As a result, there is a complex problem of analyzing the risks of objects, which is solved through the value by which the integral risk can be minimized. Risks are considered as the antipotential of the system development, which are the retarders of the reproduction rate of the system. The authors give a brief characteristic of an intellectual expansive system and propose approaches to modeling the type of functional dependence of the integral risk of functioning of such a system on the set of risks, measured, as a rule, in synthetic scales of pair comparisons. The solution to the problem of reducing the dimensionality of the influencing factors (private risks) by the vector compression method (in group and interscale formulations) is described. The paper presents an original method of processing matrices of incomplete pairwise comparisons with fuzzy information based on the idea of constructing benchmark-consistent solutions. Examples of applications of the vector compression method to solve practical problems are given. The paper presents an original method of processing matrices of incomplete pairwise comparisons with fuzzy specified information, based on the idea of constructing benchmark-consistent solutions.

In [9], the following two optimization problems on analysis of acyclic orgraphs are solved. The first one consists of determining the minimal (by volume) set of arcs whose removal from the acyclic orgraph breaks all paths passing through a subset of its vertices. The second problem is to determine the smallest set of arcs, whose introduction into the acyclic orgraph turns it into a strongly connected one. The first problem was solved by reducing it to the problem of maximal flow rate and minimal section. The second problem was solved by calculating the minimum number of input arcs and determining the smallest set of input arcs in terms of the minimum coverage of the arcs of the acyclic orgraph. The solution of these problems extends to an arbitrary orgraph by distinguishing it in the components of cyclic equivalence and the arcs between them.

The paper [10] considers the reliability function of a system consisting of k of n, under the conditions when the failures of its components lead to an increase in the load on the remaining ones and, consequently, to a change in their residual lifetime. It should be noted that the development of models is able to consider that failures of system components lead to a decrease in the residual lifetime of the remaining ones, which is of crucial importance in the tasks of increasing the reliability of the system. In [10], a new approach based on the application of order statistics of the system components' service life to model this situation is proposed. An algorithm for calculating the system reliability function and two moments of its no-failure operation time is developed. Numerical research includes sensitivity analysis for cases of the considered model based on two real systems. The obtained results show the sensitivity of system reliability characteristics to the form of service life distribution, as well as to the value of variation coefficient at a fixed average value.

Conflicts of Interest: The authors declare no conflict of interest.

References

1. Chaudhry, M.; Gai, J. Analytic and Computational Analysis of GI/Ma,b/c Queueing System. *Mathematics* **2022**, *10*, 3445. [CrossRef]
2. Tsitsiashvili, G. Processing Large Outliers in Arrays of Observations. *Mathematics* **2022**, *10*, 3399. [CrossRef]
3. Chaudhry, M.; Goswami, V. The Geo/Ga,Y/1/N Queue Revisited. *Mathematics* **2022**, *10*, 3142. [CrossRef]
4. Briš, R.; Jahoda, P. Really Ageing Systems Undergoing a Discrete Maintenance Optimization. *Mathematics* **2022**, *10*, 2865. [CrossRef]
5. Almalaq, A.; Albadran, S.; Mohamed, M. Deep Machine Learning Model-Based Cyber-Attacks Detection in Smart Power Systems. *Mathematics* **2022**, *10*, 2574. [CrossRef]
6. Seong, J. Theoretical Bounds on the Number of Tests in Noisy Threshold Group Testing Frameworks. *Mathematics* **2022**, *10*, 2508. [CrossRef]
7. Vassiliou, P. Limiting Distributions of a Non-Homogeneous Markov System in a Stochastic Environment in Continuous Time. *Mathematics* **2022**, *10*, 1214. [CrossRef]
8. Zhigirev, N.; Bochkov, A.; Kuzmina, N.; Ridley, A. Introducing a Novel Method for Smart Expansive Systems's Operation Risk Synthesis. *Mathematics* **2022**, *10*, 427. [CrossRef]
9. Tsitsiashvili, G.; Bulgakov, V. New Applied Problems in the Theory of Acyclic Digraphs. *Mathematics* **2022**, *10*, 45. [CrossRef]
10. Rykov, V.; Ivanova, N.; Kozyrev, D.; Milovanova, T. On Reliability Function of a k-out-of-n System with Decreasing Residual Lifetime of Surviving Components after Their Failures. *Mathematics* **2022**, *10*, 4243. [CrossRef]

Article

Analytic and Computational Analysis of GI/M^{a,b}/c Queueing System

Mohan Chaudhry and Jing Gai *

Department of Mathematics and Computer Science, Royal Military College of Canada, Kingston, ON K7K 7B4, Canada
* Correspondence: jing.gai@rmc.ca

Abstract: Bulk-service queueing systems have been widely applied in many areas in real life. While single-server queueing systems work in some cases, multi-servers can efficiently handle most complex applications. Bulk-service, multi-server queueing systems (compared to well-developed single-server queueing systems) are more complex and harder to deal with, especially when the inter-arrival time distributions are arbitrary. This paper deals with analytic and computational analyses of queue-length distributions for a complex bulk-service, multi-server queueing system GI/M^{a,b}/c, wherein inter-arrival times follow an arbitrary distribution, a is the quorum, and b is the capacity of each server; service times follow exponential distributions. The introduction of quorum a further increases the complexity of the model. In view of this, a two-dimensional Markov chain has to be involved. Currently, it appears that this system has not been addressed so far. An elegant analytic closed-form solution and an efficient algorithm to obtain the queue-length distributions at three different epochs, i.e., pre-arrival epoch (p.a.e.), random epoch (r.e.), and post-departure epoch (p.d.e.) are presented, when the servers are in busy and idle states, respectively.

Keywords: queues; bulk service; multi-server; Markov chain; quorum

MSC: 60-08; 60J27

Citation: Chaudhry, M.; Gai, J. Analytic and Computational Analysis of GI/M^{a,b}/c Queueing System. *Mathematics* **2022**, *10*, 3445. https://doi.org/10.3390/math10193445

Academic Editors: Gurami Tsitsiashvili and Alexander Bochkov

Received: 16 August 2022
Accepted: 8 September 2022
Published: 22 September 2022

Publisher's Note: MDPI stays neutral with regard to jurisdictional claims in published maps and institutional affiliations.

1. Introduction

Queueing theory consists of a powerful tool for modelling and analytically studying many complex systems, such as computer networks, banks, telecommunications, manufacturing, and transportation systems. Compared to well-developed single-server non-bulk queueing systems, bulk-service systems have an extensive mathematical theory. They are more complex and harder to deal with. In a bulk-service queue, a group (or batch) of customers can be served simultaneously. Examples of their applications can be seen in shuttle-bus services, freight trains, express elevators, tour operators, and batch servicing in manufacturing processes. This topic, due to its perceived applicability, has attracted the attention of many researchers over several decades. At an early stage, some simple bulk-service models, such as single-server systems GI/M^b/1 and M/M^a/1 were studied by Shyu [1] and Gross et al. [2], respectively. Neuts [3] first introduced a quorum bulk service rule to create more complex models necessary to describe certain realistic situations. He considered a queueing system with Poisson arrivals and a general service-time distribution M/G^{a,b}/1, where a is the quorum and b is the capacity of the server. Easton and Chaudhry [4] extended these results to the case where the inter-arrival times were Erlangian with the η-stage, E$_\eta$/M^{a,b}/1. Later, Chaudhry and Madill [5] gave a solution for a more general queueing system GI/M^{a,b}/1. An alternate method was given in Neuts' book [6], wherein he describes the application of his matrix geometric approach to the GI/PH^{a,b}/1 system, which has a phase-type service-time distribution. However, these systems are single-server queues. For many other variations of bulk-service queues, such as bulk service queues with vacations or bulk-service queues of the type M/G/1, one may view the survey paper written by Sasikala and Indhira [7]. In this survey, which had over 100 publications, most of the models considered were single server queues.

Multi-server queueing systems are an important class of queueing processes and have broad practical applications. However, such systems are more complex and harder to deal with compared to single-server queueing systems, especially when the inter-arrival time distribution is arbitrary. Medhi [8] investigated a queue with Poisson arrivals $M/G^{a,b}/c$, but his method was not analytically tractable for $c > 2$. Related work has also been conducted by Sim [9] on $M/M^{a,b}/c$ by using algorithmic methods but no numerical results were given. Sim [10] solved the η-phase Erlangian arrivals $E_\eta/M^{a,b}/c$ system for the random epoch probabilities in the steady state and discussed his results in the context of a transportation system. Adan and Resing [11] derived and presented the numerical results of the queue-length distributions for models $M/COXIAN\text{-}2^{a,b}/c$ and $M/E_\eta{}^{a,b}/c$. Compared to our model $GI/M^{a,b}/c$, the most relevant model studied by other researchers was $GI/M^b/c$, where the quorum was set to 1. Goswami et al. [12] solved the finite-buffer $GI/M^b/c$ model by the supplementary variable technique. Shyu [13], as well as Chaudhry and Templeton [14], dealt with the distribution of the number of customers in the system without considering the server being busy or idle. Therefore, there is no information regarding server utilization. Moreover, the numerical results for the system $GI/M^b/c$ are not available.

To make the model useful for applications, in this paper, we considered analytic and computational aspects to determine the performance of a complex bulk-service, multi-server queueing system $GI/M^{a,b}/c$. The model $GI/M^{a,b}/c$ is an extension of the system $GI/M^b/c$ (Shyu [13] as well as by Chaudhry and Templeton [14]), by introducing quorum in the multi-server system $GI/M^b/c$. A quorum refers to the minimum number of customers that are required in the waiting line before service commences, e.g., a ferry will not start until the quorum is met, or if we are dealing with transportation problems, a bus may not start until we have the quorum. This is an important policy desired by the service providers to reduce the business cost and maximize server utilization. The adding of the quorum policy makes the model closer to the real situation, but it also makes the model more complex to study. In view of this, a two-dimensional Markov chain has to be involved where the first dimension corresponds to the state of the servers (busy or idle) and the second dimension corresponds to the number of customers in the queue. We give an elegant analytic closed-form solution to obtain the queue-length distributions at three different epochs, such as pre-arrival epoch (p.a.e.), random epoch (r.e.), and post-departure epoch (p.d.e.), not only for the system in a busy state, but also in an idle state. In the case of the idle state, the probabilities were obtained by simultaneously solving the $c \times a$ equations, some of which contained infinite series, which needed to be truncated to obtain the results. Instead of truncation, which leads to approximate results, we derived a closed-form solution and proposed an efficient algorithm to fix this problem. The model $GI/M^{a,b}/c$ that we considered includes most models ([1,2,4–6,8–10,13,14]) as special cases. Our model was validated in giving numerical results with the desired degree of accuracy and trivial computational costs. By selecting particular numbers for the parameters a, b and c, and inter-arrival time distributions, the numerical results produced by our model match the ones provided in those simpler models as expected.

The paper is organized as follows. In the following section, we describe the queueing model $GI/M^{a,b}/c$, and establish a transition probability matrix (t.p.m.) for the system in Section 3. In Sections 4–6, we obtain the queue-length distributions at three different epochs, such as pre-arrival epoch (p.a.e.), random epoch (r.e.), and post-departure epoch (p.d.e.). To make the model useful for applications, sample numerical results are provided in Section 7.

2. Model Description

In this continuous-time queueing system $GI/M^{a,b}/c$, there are c independent servers, each serving at the rate μ. The customers arrive at the rate λ according to a renewal process with an arbitrary inter-arrival time distribution $A(t)$. One of the idle c servers starts the service as soon as the number of customers (including the new arriving customer) in the queue reaches quorum a. Each c server is able to serve up to b customers simultaneously. This indicates that if the server completes a service and finds less than the quorum a in the queue, it will become idle until a is reached. The service times of each server are independently–identically exponentially distributed random variables (i.i.e.d.r.v.'s). We consider the system to be in a steady state with the traffic intensity $\rho = \lambda/(bc\mu) < 1$. The queue discipline is first-come first-serve (FCFS) by batches.

3. Transition Probability Matrix (t.p.m.)

In the queueing system $GI/M^{a,b}/c$, the states occurring at the instants immediately before the arrivals form an embedded Markov chain (I.M.C.). The state seen by an arriving customer can be described by (S_n, n), where $n \geq 0$ is the queue-length and S_n is a supplementary flag defined as

$$S_n = \begin{cases} I(k), & \text{if k servers are idle,} \quad 1 \leq k \leq c, \quad 0 \leq n \leq a-1, \\ B, & \text{if all servers are busy,} \quad n \geq 0. \end{cases}$$

We define the system as busy if all the servers are busy ($S_n = B$), and idle if at least one server is idle ($S_n = I(k)$, k is the number of idle servers). The queue-length n can be written as $n = qb + n_0, 0 \leq n_0 \leq b-1$, where q is the nearest lower non-negative integer of the fraction n/b, denoting the available number of full size batches (the batch size is b) in the queue waiting for service.

To build a t.p.m. of the system, we first define the following probabilities.

1. $[l|m;t]$ and $[l|m]$, where $0 \leq l \leq m \leq c$, and there are less than a customers waiting in the queue at the beginning of the period, thus $q = 0$. Here,

$$[l|m;t] = \binom{m}{l}(1 - e^{-\mu t})^l (e^{-\mu t})^{m-l}$$

is the conditional probability that l of m servers complete services during an inter-arrival period of duration t, given that m servers are busy ($c - m$ servers are idle) at the beginning of the period. Moreover, $[l|m]$ is defined as

$$[l|m] = \int_0^\infty [l|m;t]dA(t), \qquad 0 \leq l \leq m \leq c. \tag{1}$$

2. $\{l|c;q\}$ is the conditional probability that l of c servers become idle during an inter-arrival period, given that all c servers are busy at the beginning of the period, and q ($q \geq 1$) batches of customers are waiting for the services. Assume that a time V has elapsed when the last batch of q batches enters service. In this case, the c servers have been processed at a rate of $c\mu$ until time V has elapsed. When all c servers are busy, the number of departed batches follows a Poisson process with a rate $c\mu$. The time V is Erlang-distributed, so it is the sum of q exponential random variables with a rate $c\mu$, implying that the probability density function (p.d.f.) of V is given by

$$p(v) = \frac{(c\mu)(c\mu v)^{q-1}e^{-c\mu v}}{(q-1)!}, \qquad v > 0.$$

After all the waiting q batches leave the queue, there is time $t - V$ remaining to have l batches processed. The probability that these l batches complete the service during period $t - V$ is $[l|c; t - V]$. Therefore

$$\{l|c;q\} = \int_0^\infty \int_0^t \binom{c}{l}(1 - e^{-(t-v)\mu})^l (e^{-(t-v)\mu})^{c-l} \frac{(c\mu)(c\mu v)^{q-1}e^{-c\mu v}}{(q-1)!} dv dA(t). \quad (2)$$

3. $(l|c)$ is the conditional probability that l batches complete service during an inter-arrival period of duration t, given that all the c servers are busy at the beginning of the period and still busy at the end of the period. In this case, the number of batches served in time t is distributed as a Poisson process at a rate of $c\mu$:

$$(l|c) = \int_0^\infty \frac{e^{-c\mu t}(c\mu t)^l}{l!} dA(t), \qquad l \geq 0. \quad (3)$$

Remark 1.

1. $[0|c] = (0|c) = \int_0^\infty e^{-c\mu t} dA(t) \equiv K_0$.
   Though $[0|c]$ and $(0|c)$ give identical results, they have totally different meanings. $[0|c]$ is for the case when $(c-1)$ servers are busy and $(a-1)$ customers are in queue. After one customer arrives, all the servers become busy without any departures during the inter-arrival time. In this situation, the number of customers in the queue must be zero. Moreover, $(0|c)$ is for the case that all the servers are already busy before an arrival, and no departures happen during an inter-arrival time. In this situation, the queue-length can be any non-negative number.
2. *It is easy to prove that $(l|c) = \{0|c;l\}$.*

Let J_r be the system state on the arrival of the rth customer who sees n customers in the queue. The entry of the one-step t.p.m. T from state (S_i, i) to state (S_j, j) is

$$[T_{(S_i,i),(S_j,j)}] = P(J_{r+1} = (S_j, j)|J_r = (S_i, i)), \qquad i \geq 0, j \geq 0,$$

implying that the $(r+1)$th arriving customer sees j customers waiting in the queue with the server state S_j, given that the previous rth arriving customer saw i customers waiting in the queue with the server state S_i.

The Markov chain (see Tables 1–4) for this system is two-dimensional rather than the usual one-dimensional. The t.p.m. can be formed as four sub-matrices, which are shown in Tables 1–4.

We describe the four sub-matrices that form the t.p.m.

$$T = \begin{bmatrix} \mathbf{T}_{Idle \to Idle} & \mathbf{T}_{Idle \to Busy} \\ \mathbf{T}_{Busy \to Idle} & \mathbf{T}_{Busy \to Busy} \end{bmatrix}. \quad (4)$$

(I) $\mathbf{T}_{Idle \to Idle}$. In this situation, the number of customers waiting in queue is less than a. Assume that there are k_i servers idle at the beginning of the inter-arrival time period, and k_j servers idle at the end of the inter-arrival time period, $1 \leq k_i \leq k_j \leq c$.

$$[T_{(S_i,i),(S_j,j)}] = \begin{cases} [T_{(I(k_i),i),(I(k_j),i+1)}] = [(k_j - k_i)|(c - k_i)] & \text{if } 0 \leq i < a - 1, j = i+1, \\ [T_{(I(k_i),a-1),(I(k_j),0)}] = [(k_j - k_i + 1)|(c - k_i + 1)] & \text{if } i = a - 1, j = 0. \end{cases} \quad (5)$$

(II) $\mathbf{T}_{Busy \to Idle}$. All the servers are busy at the beginning of the period, and $k(1 \leq k \leq c)$ servers are idle at the end of the period, implying that the number of customers in the queue, say j, at the end of the period, must be less than a, i.e., $j < a$. In a manner similar to what we define for $n = qb + n_0, 0 \leq n_0 \leq b - 1$, we need to arrange i customers who are waiting in queue, with FCFS discipline, into q full-size batches and a batch holding the remainders, i.e., $i = qb + i_0, 0 \leq i_0 \leq b - 1$.

$$[T_{(S_i,i),(S_j,j)}] = \begin{cases} [T_{(B,i),(I(k),i+1)}] = [k|c] & \text{if } 0 \leq i < a - 1, j = i+1, \\ [T_{(B,qb+i_0),(I(k),i_0+1)}] = \{k|c;q\} & \text{if } 0 \leq i_0 < a - 1, q \geq 1, j = i_0 + 1, \\ [T_{(B,qb+i_0),(I(k),0)}] = \{k|c;q+1\} & \text{if } a - 1 \leq i_0 \leq b - 1, q \geq 0, j = 0. \end{cases} \quad (6)$$

(III) $\mathbf{T}_{Idle \to Busy}$. The system is idle at the beginning of the time period. After one customer arrives, all the servers become busy and are still busy at the end of the time period.

This case appears only if the number of customers waiting in queue is $a - 1$, and there is only one server idle at the beginning of the time period.

$$[T_{(S_i,i),(S_j,j)}] = \begin{cases} [T_{(I(1),a-1),(B,0)}] = [0|c] & \text{if } i = j - 1, j = 0, \\ [T_{(I(k_i),i),(B,j)}] = 0 & \text{otherwise.} \end{cases} \tag{7}$$

(IV) $\mathbf{T}_{Busy \to Busy}$. All the servers are busy from the beginning to the end of the period, and the number of batches served in time t follows the Poisson process with rate $c\mu$.

$$[T_{(S_i,i),(S_j,j)}] = \begin{cases} [T_{(B,qb+i_0),(B,(q-l)b+i_0+1)}] = (l|c) & \text{if } 0 \le i_0 < b - 1, 0 \le l \le q, \\ & i = qb + i_0, j = (q-l)b + 1, \\ [T_{(B,qb+i_0),(B,0)}] = (q+1|c) & \text{if } a - 1 \le i_0 \le b - 1 \text{ and } j = 0, q \ge 0. \end{cases} \tag{8}$$

Finally, $[T_{(S_i,i),(S_j,j)}] = 0$ if $j > i + 1$ is true for all of the above I–IV cases. By using identities 1 and 2, it can be easily proven that the sum of all the entries in t.p.m. equals one.

Identity 1. $\sum_{l=1}^{c} \{l|c; q\} + \sum_{i=0}^{q} (i|c) = 1$ *for* $q > 0$. *This equation shows that the sum of all the conditional probabilities in each row of t.p.m. (when the initial system state is busy) equals one.*

Proof.

$$\sum_{l=1}^{c} \{l|c; q\} = \int_0^\infty \int_0^t \sum_{l=1}^{c} \binom{c}{l} (1 - e^{-(t-v)\mu})^l (e^{-(t-v)\mu})^{c-l} \times \frac{(c\mu)(c\mu v)^{q-1} e^{-c\mu v}}{(q-1)!} dv \, dA(t)$$

$$= \int_0^\infty \int_0^t (1 - e^{-c\mu(t-v)}) \frac{(c\mu)(c\mu v)^{q-1} e^{-c\mu v}}{(q-1)!} dv \, dA(t)$$

$$= \underbrace{\int_0^\infty \int_0^t \frac{(c\mu)(c\mu v)^{q-1} e^{-c\mu v}}{(q-1)!} dv \, dA(t)}_{\text{Term 1}} - \underbrace{\int_0^\infty \int_0^t \frac{(c\mu)(c\mu v)^{q-1} e^{-c\mu t}}{(q-1)!} dv \, dA(t)}_{\text{Term 2}}.$$

"Term 1" in the above equation can be simplified as $1 - \sum_{i=0}^{q-1} (i|c)$ by using the results that the CDF of Erlang is $1 - \sum_{i=0}^{q-1} \frac{(c\mu t)^i e^{-c\mu t}}{i!}$ and $(i|c) = \int_0^\infty \frac{(c\mu t)^i e^{-c\mu t}}{i!} dA(t)$. "Term 2" can be simplified to $(q|c)$. Combining these two terms gives $\sum_{l=1}^{c} \{l|c; q\} = 1 - \sum_{i=0}^{q} (i|c)$. $\square$

Identity 2. $\sum_{i=m}^{c} [(i - m)|(c - m)] = 1, 0 \le m \le c$. *This equation shows that, when the initial system state is idle, the sum of all the conditional probabilities in each row of t.p.m. equals one.*

Proof. $\sum_{i=m}^{c} [(i - m)|(c - m)]$
$= \int_0^\infty \sum_{i=m}^{c} \binom{c-m}{i-m} (1 - e^{-\mu t})^{i-m} (e^{-\mu t})^{c-i} dA(t)$
$= \int_0^\infty dA(t) = 1.$
$\square$

Table 1. Submatrix $\mathbf{T}_{Idle \rightarrow Idle}$.

(S_n,n)	$(I(c),0)$	$(I(c),1)$	$\cdots$	$(I(c),a-1)$	$(I(c-1),0)$	$(I(c-1),1)$	$\cdots$	$(I(c-1),a-1)$	$\cdots$	$(I(2),0)$	$(I(2),1)$	$\cdots$	$(I(2),a-1)$	$(I(1),0)$	$(I(1),1)\cdots$	$(I(1),a-1)$
$(I(c),0)$		[0|0]														
$\vdots$			$\ddots$													
$(I(c),a-2)$				[0|0]												
$(I(c),a-1)$	[1|1]				[0|1]											
$(I(c-1),0)$		[1|1]				[0|1]										
$\vdots$	$\vdots$		$\ddots$				$\ddots$									
$(I(c-1),a-2)$				[1|1]				[0|1]								
$(I(c-1),a-1)$	[2|2]				[1|2]											
$\vdots$	$\vdots$		$\ddots$			$\ddots$			$\ddots$								
$(I(2),0)$		[(c-2)|(c-2)]				[(c-1)|(c-2)]						[0|(c-2)]					
$\vdots$	$\vdots$		$\ddots$				$\ddots$		$\ddots$			$\ddots$					
$(I(2),a-2)$				[(c-2)|(c-2)]				[[c-3|c-2]						[0|(c-2)]			
$(I(2),a-1)$	[(c-1)|(c-1)]				[(c-2)|c-1)]						[1|(c-1)]				[0|(c-1)]		
$(I(1),0)$		[(c-1)|(c-1)]				[(c-2)|(c-1)]						[1|(c-1)]				[0|(c-1)]	
$\vdots$	$\vdots$		$\ddots$				$\ddots$		$\ddots$			$\ddots$					
$(I(1),a-2)$				[(c-1)|(c-1)]				[(c-2)|(c-1)]						[1|(c-1)]			[0|(c-1)]
$(I(1),a-1)$	[c|c]				[(c-1)|c]						[2|c]				[1|c]		

Table 2. Submatrix $\mathbf{T}_{Busy \to Idle}$.

(S_n, n)	$(I((c)),0)$	$(I((c)),1)$ ⋯	$(I(c),a-1)$	$(I(c-1),0)$	$(I(c-1),1)$ ⋯	$(I(c-1),a-1)$ ⋯	$(I(2),0)$	$(I(2),1)$ ⋯	$(I(2),a-1)$	$(I(1),0)$	$(I(1),1)$ ⋯	$I((1),a-1)$
$(B,0)$		[c|c]			[(c-1)|c]			[2|c]			[1|c]	
⋮	⋮	⋱			⋱	⋱		⋱			⋱	
$(B,a-2)$			[c|c]			[(c-1)|c]			[2|c]			[1|c]
$(B,a-1)$	{c|c;1}			{(c-1)|c;1}			{2|c;1}			{1|c;1}		
⋮	⋮	⋱			⋱	⋱		⋱			⋱	
$(B,b-1)$	{c|c;1}			{(c-1)|c;1}			{2|c;1}			{1|c;1}		
(B,b)		{c|c;1}			{(c-1)|c;1}			{2|c;1}			{1|c;1}	
⋮	⋮	⋱			⋱	⋱		⋱			⋱	
$(B,b+a-2)$			{c|c;1}			{(c-1)|c;1}			{2|c;1}			{1|c;1}
$(B,b+a-1)$	{c|c;2}			{(c-1)|c;2}			{2|c;2}			{1|c;2}		
⋮	⋮	⋱			⋱	⋱		⋱			⋱	
$(B,2b-1)$	{c|c;2}			{(c-1)|c;2}			{2|c;2}			{1|c;2}		
$(B,2b)$		{c|c;2}			{(c-1)|c;2}			{2|c;2}			{1|c;2}	
⋮	⋮	⋱			⋱	⋱		⋱			⋱	
$(B,(q-1)b)$		{c|c;q-1}			{(c-1)|c;q-1}			{2|c;q-1}			{1|c;q-1}	
⋮	⋮	⋱			⋱	⋱		⋱			⋱	
$(B,(q-1)b+a-2)$			{c|c;q-1}			{(c-1)|c;q-1}			{2|c;q-1}]			{1|c;q-1}
$(B,(q-1)b+a-1)$	{c|c;q}			{(c-1)|c;q}			{2|c;q}			{1|c;q}		
⋮	⋮	⋱			⋱	⋱		⋱			⋱	
$(B,qb-1)$	{c|c;q}			{(c-1)|c;q}			{2|c;q}			{1|c;q}		
(B,qb)		{c|c;q}			{(c-1)|c;q}			{2|c;q}			{1|c;q}	
⋮	⋮	⋱			⋱	⋱		⋱			⋱	

Table 3. Submatrix $\mathbf{T}_{Busy \to Busy}$.

(S_n, n)	$(B,0)$	$(B,1)$	$(B,2)$	$\cdots$	(B,a)	$\cdots$	(B,b)	$\cdots$
$(B,0)$		(0 | c)						
$(B,1)$			(0 | c)					
⋮	⋮			⋱		⋱		⋱
$(B,a-2)$								
$(B,a-1)$	(1 | c)				(0 | c)			
⋮	⋮			⋱		⋱		⋱
$(B,b-1)$	(1 | c)						(0 | c)	
(B,b)		(1 | c)						
⋮	⋮			⋱		⋱		⋱
$(B,b+a-2)$								
$(B,b+a-1)$	(2 | c)							
⋮	⋮			⋱		⋱		⋱
$(B,2b-1)$	(2 | c)							
$(B,2b)$		(2 | c)						
⋮	⋮			⋱		⋱		⋱
$(B,(q-1)b)$		((q-1) | c)						
⋮	⋮			⋱		⋱		⋱
$(B,(q-1)b+a-2)$								
$(B,(q-1)b+a-1)$	(q | c)				((q-1) | c)			
⋮	⋮			⋱		⋱		⋱
$(B,qb-1)$	(q | c)						((q-1) | c)	
(B,qb)		(q | c)						
⋮	⋮			⋱		⋱		⋱

Since the Markov chain under consideration is irreducible, positive recurrent and aperiodic, it has a limiting distribution if and only if $\rho = \lambda/bc\mu < 1$. In view of this, $\lim_{r \to \infty} P(J_r = (S_n, n)) = X(S_n, n)$ exists. In this case, the limiting distribution is given by $\mathbf{X} = \mathbf{XT}$ where $\mathbf{T}$ is t.p.m. defined in (4), and the vector $\mathbf{X}$ has the form

$$\mathbf{X} = [X(I(c),0), \cdots, X(I(c), a-1), \cdots, X(I(1),0), \cdots, \tag{9}$$
$$X(I(1), a-1), X(B,0), \cdots X(B,1), \cdots],$$

where $X(I(k), n), 0 \leq n < a$ and $X(B,n), n \geq 0$, respectively, denote the p.a.e. unnormalized probabilities that an arriving customer sees n customers in queue, k of c servers idle, and n customers in queue, with all servers busy. If such a vector $\mathbf{X}$ exists, it will be the vector of the steady state p.a.e. probabilities up to some normalizing constant.

Table 4. Submatrix $\mathbf{T}_{Idle \to Busy}$.

(S_n, n)	$(B,0)$	$(B,1)$	$\cdots$	$(B,a-1)$	$\cdots$	(B,b)	$\cdots$
$(I(c),0)$							
$\vdots$							
$(I(c),a-2)$							
$(I(c),a-1)$							
$(I(c-1),0)$							
$\vdots$							
$(I(c-1),a-2)$							
$(I(c-1),a-1)$							
$\vdots$							
$(I(2),0)$							
$\vdots$							
$(I(2),a-2)$							
$(I(2),a-1)$							
$(I(1),0)$							
$\vdots$							
$(I(1),a-2)$							
$(I(1),a-1)$	$[0\,|\,c]$						

4. Queue-Length Distributions at Pre-Arrival Epoch

4.1. The Busy Server Probabilities

When all the servers are busy during an inter-arrival time period, for the queueing model $GI/M^{a,b}/c$, the service times for batches are i.i.d.r.v.$'$ s, having exponential distributions. Thus, the number of batches that complete service during an arbitrary inter-arrival time will have a Poisson distribution, which implies that the probability of l service completions during an inter-arrival time A is $(l|c)$, and the probability generating function (p.g.f.) of $(l|c)$ is

$$D(z) = \sum_{l=0}^{\infty} (l|c) z^l = \bar{a}(c\mu(1-z)), \tag{10}$$

where $\bar{a}(\alpha)$ is the Laplace–Stieltjes transform (L.-S.T.) of $A(t)$, i.e., $\bar{a}(\alpha) = \int_0^{\infty} \exp(-\alpha t) dA(t)$ and

$$K_0 = \bar{a}(c\mu) = \int_0^{\infty} \exp(-c\mu t) dA(t). \tag{11}$$

Theorem 1. *For the queueing system $GI/M^{a,b}/c$, in the steady state case, the busy-server probabilities of queue length at pre-arrival epoch are given by $P^-(B,n) = X(B,n)/C_N = w^n/C_N, n \geq 0$, where w is a real root inside the unit circle of equation $D(z^b) = z = \bar{a}(c\mu(1-z^b))$ and C_N is a normalizing constant given by $C_N = \sum_{j=1}^{c} \sum_{i=0}^{a-1} X(I(j),i) + \frac{1}{1-w}$.*

Proof. When the system is busy and n customers are waiting in the queue, it is evident from t.p.m. that

$$X(B,n) = \sum_{j=0}^{\infty} (j|c) X(B, jb+n-1), \qquad n > 0. \tag{12}$$

To solve the difference Equation (12), in the same manner as by Chaudhry and Madill [5], a solution of the form $X(B,n) = z^n \ (z \neq 0), n \geq 1$ is assumed. For more details on this method, one may see Chaudhry and Templeton ([14], page 350). Substituting $X(B,n) = z^n$ into Equation (12), we have

$$z^n = \sum_{j=0}^{\infty} z^{jb+n-1}(j|c) = z^{n-1} \sum_{j=0}^{\infty} (j|c)z^{jb} = z^{n-1}D(z^b). \tag{13}$$

Combining this with Equation (10), and simplifying, we obtain the root equation

$$D(z^b) = z = \bar{a}(c\mu(1 - z^b)). \tag{14}$$

By Rouché's theorem, it can be shown that Equation (14) has a real root w inside the unit circle if $\rho = \frac{\lambda}{bc\mu} < 1$. Once the root w is found, $X(B,1), X(B,2), \cdots$ can be obtained by using $X(B,n) = w^n, n \geq 1$.

Next, we can solve for $X(B,0)$ from Equation (12) by setting $n = 1$,

$$X(B,1) = w = \sum_{j=0}^{\infty} (j|c)X(B,jb),$$

implying

$$w = (0|c)X(B,0) + (1|c)w^b + (2|c)w^{2b} + \cdots = D(w^b). \tag{15}$$

Combining (10), (14), and (15), we conclude $X(B,0) = 1$. This implies that the assumption $X(B,n) = z^n$ is true even for $n = 0$.

Finally, $P^-(B,n)$ can be obtained as the normalized $X(B,n)$ by dividing a normalizing constant C_N (see Equations (23) and (24)). $\quad\square$

4.2. The Idle Server Probabilities

The idle server unnormalized probabilities $X(I(c),0), \cdots, X(I(c), a-1), \cdots, X(I(1),0),$ $\cdots, X(I(1), a-1)$ can be obtained by $c \times a$ linear equations generated from the t.p.m. In fact, there are "$c \times a + 1$" equations, with (as usual) one being redundant.

These "$c \times a + 1$" equations are

$$X(B,0) = X(I(1), a-1)[0|c] + \sum_{i=1}^{\infty} (i|c) \sum_{l=a}^{b} X(B, (i-1)b+l-1), \tag{16}$$

$$X(I(k),j) = \sum_{m=1}^{k} X(I(m), j-1)[(k-m)|(c-m)] + X(B, j-1)[k|c] +$$
$$\sum_{i=1}^{\infty} X(B, ib+j-1)\{k|c;i\}, \tag{17}$$

and

$$X(I(k),0) = \sum_{m=1}^{k+1} X(I(m), a-1)[(k-m+1)|(c-m+1)] +$$
$$\sum_{i=1}^{\infty} \{k|c;i\} \sum_{l=a}^{b} X(B, (i-1)b+l-1), \tag{18}$$

where $1 \leq j \leq a-1, 1 \leq k \leq c$ and $X(I(c+1), a-1) = 0$.

Remark 2. *The $c \times a$ idle server unknown probabilities (unnormalized)*

$$[X(I(c),0), \cdots, X(I(c), a-1), \cdots, X(I(1),0), \cdots, X(I(1), a-1)]$$

can be obtained simultaneously by using the above $c \times a$ equations. However, large values of c or a may cause a computational problem, since the last terms in both Equations (17) and (18) are infinite series related to complex double integrals $\{k|c;i\}$ (defined in Equation (2)). In general, when we operate on an infinite series without a closed form, the series has to be truncated. Therefore, the result is approximated as we lose the tails due to this truncation. To fix these problems, we want to *simplify Equations (17) and (18) by deriving a closed form for these series. Before we move on, we need to prove the following two lemmas.*

Lemma 1. *The last term in Equation (16)*

$$\sum_{i=1}^{\infty} (i|c) \sum_{l=a}^{b} X(B, (i-1)b + l - 1) = \frac{w^{a-b-1} - 1}{1 - w}(w - K_0).$$

Proof.

$$\sum_{i=1}^{\infty} (i|c) \sum_{l=a}^{b} X(B, (i-1)b + l - 1)$$

$$= w^{a-b-1} \sum_{i=1}^{\infty} (i|c) \sum_{k=0}^{b-a} w^{ib+k} = w^{a-b-1}\frac{1 - w^{b-a+1}}{1 - w} \sum_{i=1}^{\infty} (i|c)w^{ib}$$

$$= w^{a-b-1}\frac{1 - w^{b-a+1}}{1 - w}(D(w^b) - K_0) = \frac{w^{a-b-1} - 1}{1 - w}(w - K_0)$$

by using $(0|c) = K_0$, and Equation (13). $\quad\square$

Lemma 2. *Define $J(k) = \sum_{i=1}^{\infty} w^{ib}\{k|c;i\}$, and*

$$J(k) = c\mu w^b \int_0^{\infty} \int_0^t \binom{c}{k}(1 - e^{-(t-v)\mu})^k (e^{-(t-v)\mu})^{c-k} e^{-c\mu v(1-w^b)} dv\,dA(t). \tag{19}$$

Proof.

$$\sum_{i=1}^{\infty} w^{ib}\{k|c;i\}$$

$$= \sum_{i=1}^{\infty} w^{ib} \int_0^{\infty} \int_0^t \binom{c}{k}(1 - e^{-(t-v)\mu})^k (e^{-(t-v)\mu})^{c-k} \frac{(c\mu)(c\mu v)^{i-1}e^{-c\mu v}}{(i-1)!} dv\,dA(t)$$

$$= \int_0^{\infty} \int_0^t \binom{c}{k}(1 - e^{-(t-v)\mu})^k (e^{-(t-v)\mu})^{c-k} \sum_{i=1}^{\infty} w^{ib} \frac{(c\mu)(c\mu v)^{i-1}e^{-c\mu v}}{(i-1)!} dv\,dA(t)$$

$$= c\mu w^b \int_0^{\infty} \int_0^t \binom{c}{k}(1 - e^{-(t-v)\mu})^k (e^{-(t-v)\mu})^{c-k} e^{-c\mu v(1-w^b)} \underbrace{\sum_{i=1}^{\infty} \frac{(c\mu v w^b)^{i-1}e^{-c\mu v w^b}}{(i-1)!}}_{=1,\text{ Poisson p.m.f}} dv\,dA(t).$$

$\square$

Theorem 2. *For the queueing system $GI/M^{a,b}/c$, in the steady state case, the idle server probabilities of queue length at the pre-arrival epoch are given by $P^{-}(I(k), n) = X(I(k), n)/C_N, 0 \le n < a - 1, 1 \le k \le c$, where C_N is a normalizing constant given by $C_N = \sum_{j=1}^{c} \sum_{i=0}^{a-1} X(I(j), i) + \frac{1}{1-w}$ and $X(I(k), n)$ satisfy the following equations:*

$$(i)\quad X(I(1), a-1) = \frac{1}{(1-w)K_0}(1 - w^{a-b} + K_0 w^{a-b-1} - K_0), \tag{20}$$

$$(ii)\quad X(I(k), j) = \sum_{m=1}^{k} X(I(m), j-1)[(k-m)|(c-m)] + w^{j-1}([k|c] + J(k)), 1 < j < a - 1, \tag{21}$$

$$(iii)\quad X(I(k), 0) = \sum_{m=1}^{k+1} X(I(m), a-1)[(k-m+1)|(c-m+1)] + \frac{w^{a-b-1} - 1}{1 - w}J(k). \tag{22}$$

Proof. (i) Using Lemma 1 and $[0|c] = K_0$, we can rewrite Equation (16) and directly solve for $X(I(1), a-1)$.

(ii) and (iii) Using Theorem 1, replacing $X(B, j-1)$ by w^{j-1}, $X(B, ib+j-1)$ by w^{ib+j-1}, and $X(B, (i-1)b+l-1)$ by $w^{(i-1)b+l-1}$, then applying the result of Lemma 2, we can rewrite Equations (17) and (18) as Equations (21) and (22), respectively.

We first solved $X(I(1), a-1)$ using Equation (20), and then solved other idle server probabilities recursively by using Equations (21) and (22). For more details on this, see the algorithm developed in Appendix A. $\square$

Finally, we obtained all queue-length probabilities, and needed to normalize the vector

$$\mathbf{X} = [X(I(c), 0), \cdots, X(I(c), a-1), \cdots, X(I(1), 0), \cdots,$$
$$X(I(1), a-1), X(B, 0), \cdots X(B, 1), \cdots].$$

by dividing a normalizing constant C_N, which is given by

$$C_N = \sum_{j=1}^{c} \sum_{i=0}^{a-1} X(I(j), i) + \sum_{i=0}^{\infty} X(B, i) = \sum_{j=1}^{c} \sum_{i=0}^{a-1} X(I(j), i) + \frac{1}{1-w}. \tag{23}$$

Define $\mathbf{P}^-$ as the vector of normalized p.a.e. such that

$$\mathbf{P}^- = \frac{\mathbf{X}}{C_N}. \tag{24}$$

Further, $P^-(I(k), n)$ and $P^-(B, n)$, respectively, are normalized p.a.e. probabilities and represent that k of the c servers are idle, $0 \le n < a-1$, and all servers are busy, $n \ge 0$.

4.3. Special Cases

4.3.1. Single-Server Probabilities for GI/M^{a,b}/1

The system GI/M^{a,b}/1 is a special case of GI/M^{a,b}/c when $c = 1$.

(A) When $c = 1$, the root Equation (14) is simplified to $D(z) = z = \bar{a}(\mu(1-z))$, which agrees with the root equation in the work by Chaudhry and Madill [5]; consequently, the same results of $X(B, 0), \cdots, X(B, 1), \cdots, X(B, M)$ can be obtained.

(B) Moreover, $k = m = c = 1$, $[0|0] = 1$, $[1|1] = 1 - [0|1] = 1 - K_0$, and $\sum_{i=1}^{\infty} w^{ib}\{1|1; i\} = \frac{1}{(1-w^b)}(w^b - w + (1-w^b)K_0)$. Equation (21) can be simplified to

$$X(I(1), j) = X(I(1), j-1) + w^{j-1}(1 - K_0 + \frac{1}{(1-w^b)}(w^b - w + (1-w^b)K_0))$$
$$= X(I(1), j-1) + w^{j-1}\frac{1-w}{1-w^b}.$$

This agrees with the equation in Chaudhry and Madill [5] for solving the idle server probabilities.

4.3.2. Multi-Server Queueing System GI/M^b/c

The system GI/M^b/c is a special case of GI/M^{a,b}/c when $a = 1$.

In GI/M^b/c, the system is idle only if there is no customer waiting in queue. Instead of evaluating the queue-length distributions, Chaudhry and Templeton [14] consider the distribution for the number of customers in the system for GI/M^b/c without considering the server being busy or idle. The numerical results for the system GI/M^b/c are also not available. We can see that our model includes this model as a special case, it not only produces the numerical solutions for the queue-length distributions, but also the information of the server utilization.

5. Queue-Length Distributions at Random Epoch

We are now interested in knowing the probability that the system will be in a given state at a random epoch (r.e.) in time. A random epoch is said to occur at the end of a random period of time, R, since the last p.a.e. From renewal theory, the probability associated with R, $dR(t)$ is given by $dR(t) = \lambda(1 - A(t))dt, t > 0$ (see Chaudhry and Templeton [14]). Proceeding in a manner directly analogous to that used for developing $(l|c)$, $[l|m]$ and $\{l|c;q\}$, where the services are considered during the inter-arrival time A (see Equations (1)–(3)), $(l|c)_R$, $[l|m]_R$ and $\{l|c;q\}_R$ are defined as the probabilities that such services take place during time R. The p.g.f. of $(l|c)_R$ (see proof in Appendix B) is

$$D_R(z) = \sum_{l=0}^{\infty} (l|c)_R z^l = \frac{\rho b}{1 - z}[1 - \bar{a}(c\mu(1 - z))], \tag{25}$$

and

$$(0|c)_R = [0|c]_R = \lambda \int_0^{\infty} \exp(-c\mu t)(1 - A(t))dt = \rho b(1 - K_0). \tag{26}$$

Similar to the definition for the p.a.e probability vector $\mathbf{P}^-$ in Equation (24), we define $\mathbf{P}$ as the vector of the r.e. probabilities, such that

$$\mathbf{P} = [P(I(c),0), \cdots, P(I(c), a-1), \cdots, P(I(1),0), \cdots, P(I(1), a-1), P(B,0), \cdots P(B,1), \cdots],$$

where $P(I(k),n), 0 \le n < a$ and $P(B,n), n \ge 0$, respectively, denote the r.e. probabilities that, at the end of a random period of time R after arrival, k of the c servers are idle, $0 \le n < a-1$ customers are in the queue , and all servers are busy, $n \ge 0$ customers are in the queue. The forms of the t.p.m. $\mathbf{T}$ in Tables 1–4 contain all of the information required on transitions within the queueing system in a period measured from the last p.a.e. The nature of the entries in the t.p.m. serve to indicate the probabilities associated with the transitions. Thus, if the limiting distribution is $\mathbf{P}^- = \mathbf{P}^-\mathbf{T}$ when the timeframe is the inter-arrival time, A, instead of the entries $(l|c)$, $[l|m]$ and $\{l|c;q\}$, the entries $(l|c)_R$, $[l|m]_R$ and $\{l|c;q\}_R$ are used with the timeframe, R, and $\mathbf{P} = \mathbf{P}^-\mathbf{T_R}$, where the newly formed t.p.m. $\mathbf{T_R}$ describes how the steady-state p.a.e. probabilities are transformed into steady-state probabilities for the system at a random epoch after the last p.a.e.

Remark 3. *Similar to those in the p.a.e. systems, it can be proven that the following three equations still hold for the case of r.e. systems:*

- $[0|c]_R = (0|c)_R \equiv \rho b(1 - K_0)$ *(see Equation (26));*
- $\sum_{l=1}^{c} \{l|c;q\}_R + \sum_{i=0}^{q} (i|c)_R = 1$ *for $q > 0$; and*
- $\sum_{i=m}^{c} [(i - m)|(c - m)]_R = 1, 0 \le m \le c$.

  Thus, the sum of entries in each row of t.p.m. $\mathbf{T_R}$ equals one.

5.1. The Busy-Server Probabilities

The busy-server r.e. probabilities $P(B,n), n \ge 0$ can be calculated in a similar manner as the queue-length distributions at the pre-arrival epoch described in Section 4.1. Here, we derive the closed-form busy-server probability distribution of the queue length at a random epoch. The probabilities $P(B,n), n \ge 0$ can be obtained using Equations (27) and (28) (see below). Since both are in terms of the root w, the calculations become extremely simple. The key idea to derive these two equations is based on the relations between two probabilities: $P(B,n)$ and $P^-(B,n), n \ge 0$.

Theorem 3. *For the queueing system GI/M^{a,b}/c, in the steady state case, the busy-server probabilities of the queue length at the random epoch are given by*

(i) $\quad P(B,n) = \dfrac{1}{C_N} \dfrac{\rho b(1-w)w^{n-1}}{1-w^b}, \qquad n > 0.$

(ii) $\quad P(B,0) = \dfrac{\rho b(1-K_0)}{C_N(1-w)K_0}\left(1 - w^{a-b}\right) + \dfrac{\rho b(w^{a-b-1}-1)}{C_N(1-w^b)}.$

Proof. (i) At the end of a random period of time R after arrival, if all servers are busy and the waiting line is not empty ($n > 0$), then the sizes for those batches that were taken into service during time R must be maximum ($= b$, full batch size). Since the queue length at a pre-arrival epoch will be $n - 1 + mb, m \geq 0$, it leads to r.e. probabilities as

$$P(B, n) = \sum_{m=0}^{\infty} (m|c)_R P^-(B, mb + n - 1), n > 0.$$

By using the fact that $P^-(B, n) = w^n$, and Equations (14) and (25), we have

$$
\begin{aligned}
P(B, n) &= \frac{1}{C_N} \sum_{m=0}^{\infty} (m|c)_R w^{mb+n-1} \\
&= \frac{1}{C_N} \frac{\rho b (1 - w) w^{n-1}}{1 - w^b}, \qquad n > 0.
\end{aligned}
\tag{27}
$$

(ii) In this situation, the queue length is empty at a random time while all the servers are busy, then the size for the last batch into service can be any number between $[a, b]$, and the servers at the moment when the last customer arrives are either all busy or one idle. Combining all of these possibilities, using Equation (20) and the following equation

$$
\begin{aligned}
\sum_{i=1}^{\infty} (i|c)_R \sum_{j=a}^{b} P^-(B, (i-1)b + j - 1) &= \frac{w^{a-b-1} - 1}{C_N(1 - w)} \left[\sum_{i=0}^{\infty} (i|c)_R w^{ib} - (0|c)_R \right] \\
&= \frac{w^{a-b-1} - 1}{C_N(1 - w)} \left(\frac{\rho b (1 - w)}{1 - w^b} - \rho b (1 - K_0) \right),
\end{aligned}
$$

$P(B, 0)$ can be expressed as

$$
\begin{aligned}
&P(B, 0) \\
&= (0|c)_R P^-(I(1), a - 1) + \sum_{i=1}^{\infty} (i|c)_R \sum_{j=a}^{b} P^-(B, (i-1)b + j - 1) \\
&= \frac{\rho b (1 - K_0)}{C_N(1 - w) K_0} (1 - w^{a-b} + K_0 w^{a-b-1} - K_0) + \frac{w^{a-b-1} - 1}{C_N(1 - w)} \left(\frac{\rho b (1 - w)}{1 - w^b} - \rho b (1 - K_0) \right) \\
&= \frac{\rho b (1 - K_0)}{C_N(1 - w) K_0} (1 - w^{a-b}) + \frac{\rho b (w^{a-b-1} - 1)}{C_N(1 - w^b)}.
\end{aligned}
\tag{28}
$$

$\square$

At the end of a random period of time R after arrival, if all servers are busy, the queue length n ($n \geq 0$) distribution can be evaluated by using Equations (27) and (28). In this case, both the results are in closed-form in terms of the root w.

5.2. The Idle Server Probabilities

Corollary 1. *The idle server r.e. probabilities $P(I(k), n), 0 \leq n < a$ can be obtained by using Theorem 2. The Equations (30) and (31) (see below) are modified from Equations (21) and (22) in Theorem 2 by replacing the term $[l|m]$ with $[l|m]_R$, and normalizing the probabilities from $X(I(m), j - 1)$ to $P^-(I(m), j - 1), 1 < j < a$. Moreover, we redefine $J_R(k)$ as*

$$
\begin{aligned}
J_R(k) &= \sum_{i=1}^{\infty} w^{ib} \{k|c; i\}_R \\
&= c\lambda\mu w^b \left(\int_0^{\infty} \int_0^t \binom{c}{k} (1 - e^{-(t-v)\mu})^k (e^{-(t-v)\mu})^{c-k} e^{-c\mu v(1 - w^b)} (1 - A(t)) dv dt \right)
\end{aligned}
\tag{29}
$$

$$
\text{Then, } P(I(k), 0) = \sum_{m=1}^{k+1} P^-(I(m), a - 1)[k - m + 1|c - m + 1]_R + \frac{w^{a-b-1} - 1}{1 - w} J_R(k),
\tag{30}
$$

$$
P(I(k), j) = \sum_{m=1}^{k} P^-(I(m), j - 1)[k - m|c - m]_R + w^{j-1} ([k|c]_R + J_R(k)),
\tag{31}
$$

where $1 \leq j \leq a-1, 1 \leq k \leq c, \ P^-(I(c+1), a-1) = 0.$

5.3. The Special Case: $E_\eta/M^{a,b}/c$ Queue

The system $E_\eta/M^{a,b}/1$ is a special case of $GI/M^{a,b}/c$ when the inter-arrival time is Erlang (with η phase)-distributed. Then the root Equation (14) can be simplified to

$$\left(\frac{\eta\rho b}{\eta\rho b + 1 - z^b}\right)^\eta - z = 0.$$

By replacing $dA(t)$ with $\frac{(\lambda\eta)^\eta t^{\eta-1} e^{-\lambda\eta t}}{(\eta-1)!} dt$, we can calculate p.a.e. probability distributions for both busy and idle servers by using the algorithm introduced in Appendix A. Then the r.e. probability distributions can be obtained by using Equations (27)–(31).

Sim [10] solved the η-phase Erlangian arrivals system $E_\eta/M^{a,b}/c$ only for the probabilities at r.e. and discussed the results in the context of transportation systems. Our algorithms can not only solve the systems with general inter-arrival time distributions, but also provide the solutions at different epochs. Our numerical results agree with those provided by Sim [10].

6. Queue-Length Distributions at Post-Departure Epoch

In this section, we derive the probabilities for the state of the system immediately after a real service completion takes place. It was assumed that no time elapsed after the server completed a batch before accepting a quorum-complete batch from the queue. Thus, the post-departure epoch (p.d.e.) occurred immediately after a server had either reduced the queue or became idle.

To find the p.d.e. probabilities, we need to first define an epoch—a pre-service completion epoch (p.s.e.), i.e., the instant in the time immediately before a real departure occurs (before a real service completes). Then, $P^{S-}(I(k), n)$ and $P^{S-}(B, n), n \geq 0, 1 \leq k \leq c$, respectively, are defined as the probabilities at p.s.e., when there are n customers in queue, k of c servers idle, and n customers in queue, all servers busy. It is apparent that $P^{S-}(I(c), n) = 0$ for any n.

Similarly, we define $P^+(I(k), n), 0 \leq n < a, 1 \leq k \leq c$ and $P^+(B, n), n \geq 0$, as the probabilities of the queue length at a p.d.e.

Conjecture 1. *The following relationships between p.d.e. and p.s.e. probabilities apply*

$$P^+(I(k), n) = P^{S-}(I(k-1), n), \quad 0 \leq n \leq a-1, 2 \leq k \leq c$$
$$P^+(I(1), n) = P^{S-}(B, n), \quad 0 \leq n \leq a-1,$$

$$\tag{32}$$

and

$$P^+(B, n) = P^{S-}(B, n+b), \quad n \geq 1,$$
$$P^+(B, 0) = \sum_{n=a}^{b} P^{S-}(B, n).$$

$$\tag{33}$$

Corollary 2. $P^{S-}(I(k), n), 0 \leq n < a, \ 1 \leq k \leq c$ *and* $P^{S-}(B, n), n \geq 0$ *satisfy the following equations:*

$$P^{S-}(I(k), n) = \frac{P(I(k), n)}{1 - \sum_{i=0}^{a-1} P(I(c), i)}, \quad 0 \leq n \leq a-1, 1 \leq k \leq c-1,$$
$$P^{S-}(B, n) = \frac{P(B, n)}{1 - \sum_{i=0}^{a-1} P(I(c), i)}, \quad n \geq 0.$$

$$\tag{34}$$

Proof. When the service time distribution is exponential, service completions, real or potential, occur at random epochs. The probabilities, $P^{S-}(I(k), n), 0 \leq n < a, 1 \leq k \leq c$ and $P^{S-}(B, n), n \geq 0$ can be found by conditioning the r.e. probabilities to ensure that at least one server is busy. Thus, using the results of r.e. probabilities given in Theorem 3, we can obtain p.d.e. probabilities for both busy and idle servers from Equations (32)–(34). $\quad\square$

Remark 4.

(i)   *When we set $c = 1$, these probabilities agree with those of Chaudhry and Madill [5] for the system GI/M^{a,b}/1.*

(ii) *As a check on the algebra, also useful as a computational check, we note that, using (32)–(34),*

$$\sum_{k=1}^{c}\sum_{j=0}^{a-1} P^+(I(k), j) + \sum_{j=0}^{\infty} P^+(B, j) = \frac{\sum_{k=1}^{c-1}\sum_{j=0}^{a-1} P(I(k), j) + \sum_{j=0}^{\infty} P(B, j)}{1 - \sum_{i=0}^{a-1} P(I(c), i)}$$

$$= \frac{1 - \sum_{i=0}^{a-1} P(I(c), i)}{1 - \sum_{i=0}^{a-1} P(I(c), i)} = 1,$$

as it should be.

7. Numerical Results

In this section, we present some numerical results for various inter-arrival time distributions such as η-phase Erlang (E$_\eta$), deterministic (D), and uniform (U). All the examples we considered have the same mean value of the inter-arrival time $E(A) = 1/\lambda$. The root equation (see Equation (14)), probability density functions (p.d.f.) of inter-arrival time A, and p.d.f. of a random period time R for these three distributions are summarized in Table 5.

Table 5. Root Equations, p.d.f.s of $A(t)$, $R(t)$, and mean value of of $A(t)$ for E$_\eta$/M^{a,b}/c, D/M^{a,b}/c and U/M^{a,b}/c.

Inter-arrival time distributions	Root Equations (Equation (14))	p.d.f. of $A(t)$	p.d.f. of $R(t)$	$E(A)$
η-phase Erlang	$\left(\dfrac{\eta\rho b}{\eta\rho b+1-z^b}\right)^{\eta} - z = 0$	$\dfrac{(\lambda\eta)^{\eta} t^{\eta-1} exp(-\lambda\eta t)}{(\eta-1)!}$	$\lambda \sum_{n=0}^{\eta-1} \dfrac{(\lambda\eta t)^n exp(-\lambda\eta t)}{n!}$	$1/\lambda$
Deterministic	$exp\left(-\dfrac{1-z^b}{\rho b}\right) - z = 0$	$\delta(t - 1/\lambda)$	$\begin{cases} \lambda, & if \ \ t < \frac{1}{\lambda} \\ 0, & if \ \ t \geq \frac{1}{\lambda} \end{cases}$	$1/\lambda$
Uniform	$\dfrac{exp\left(-\frac{1-z^b}{\rho b}\right)}{\varphi c\mu(1-z^b)} \times [exp(\varphi c\mu(1 - z^b)/2) - exp(-\varphi c\mu(1 - z^b))/2] - z = 0, \ \varphi = t_2 - t_1,$ is the interval width	$1/\varphi$	$\begin{cases} \lambda, & if \ \ t < t_1 \\ \frac{1}{\varphi} + \frac{\lambda}{2} - \frac{\lambda t}{\varphi}, & if \ \ t_1 \leq t < t_2 \\ 0, & if \ \ t \geq t_2 \end{cases}$ $t_1 = \frac{1}{\lambda} - \frac{\varphi}{2}, \ t_2 = \frac{1}{\lambda} + \frac{\varphi}{2}$	$1/\lambda$

Besides the calculations for the queue-length probabilities at the pre-arrival, random, and post-departure epochs for both idle and busy systems, we also considered the performance measures, such as the mean (denoted as LQ^e) and the standard deviations (denoted as SDLQ^e) of the queue length; the mean (denoted as $E^e[I(k)]$) and variance (denoted as $Var^e[I(k)]$) of the idle servers. The symbol "e" denotes the epoch state, which can be pre-arrival (e = "–"), random (e = " "), or post-departure (e = "+"). We define $PB^e = \sum_{n=0}^{\infty} P^e(B, n)$ as the probability that an arriving customer sees the system busy at e epoch, and $PI^e = \sum_{n=0}^{a-1}\sum_{k=1}^{c} P^e(I(k), n)$ is the probability that the system is idle at e epoch. The probabilities of the queue length at three different epochs are presented in closed form. Since most of these probabilities are irrational, for computational purposes, we need to set the precision ε. Throughout all computations in the following examples, we use $\varepsilon = 10^{-20}$ as the precision. Due to the rounding error, the sum of the probabilities may not be one.

The results of the $E_6/M^{5,10}/5$ queue with traffic intensities $\rho = 0.1, 0.5, 0.9$ for both busy and idle servers at pre-arrival epoch are presented in Tables 6 and 7, respectively. When we set the number of servers to 1, our results match with those obtained for $E_6/M^{5,10}/1$ by Chaudhry et al. [5].

We considered three systems $E_6/M^{a,10}/5$, $D/M^{a,10}/5$, and $U/M^{a,10}/5$ ($t_1 = 0.875/\lambda$, $t_2 = 1.125/\lambda$, $\varphi = 0.25/\lambda$). All three systems have the same mean value of inter-arrival time $E(A) = 1/\lambda$. In Table 8, we present the performance measures for these three systems for idle servers at three different epochs with varied $a = 1, 4, 7$ and $\rho = 0.1, 0.5, 0.9$. In Figure 1, we compare the performance of $D/M^{4,10}/5$ for busy servers at pre-arrival epochs with $\rho = 0.1, 0.3, 0.5, 0.7$ and 0.9. In Figure 2, we compare the performance of $U/M^{a,10}/5$ for busy servers at pre-arrival epochs with $a = 1, 4, 7$.

Table 6. Distribution of queue lengths at pre-arrival epochs for the busy system $E_6/M^{5,10}/5$, $\rho = 0.1, 0.5, 0.9, \varepsilon = 10^{-20}$.

n	$P^-(B,n)$		
	$\rho = 0.1$	$\rho = 0.5$	$\rho = 0.9$
0	1.12×10^{-5}	0.0715625	0.0198544
1	4.45×10^{-6}	0.0612334	0.0194400
2	1.76×10^{-6}	0.0523951	0.0190343
3	7.00×10^{-7}	0.0448325	0.0186371
4	2.77×10^{-7}	0.0383615	0.0182481
5	1.10×10^{-7}	0.0328244	0.0178673
$\vdots$	$\vdots$	$\vdots$	$\vdots$
10	1.08×10^{-9}	0.015056	0.0160790
$\vdots$	$\vdots$	$\vdots$	$\vdots$
50	9.28×10^{-26}	2.95×10^{-5}	0.0069163
$\vdots$	$\vdots$	$\vdots$	$\vdots$
296	$\vdots$	6.55×10^{-22}	3.86×10^{-5}
$\vdots$	$\vdots$	$\vdots$	$\vdots$
2184	$\vdots$	$\vdots$	1.96×10^{-22}
PB^-	0.0000186	0.4957989	0.9513294
PI^-	0.9999815	0.5042011	0.0486706
SUM	1.0000001	1.0000000	1.0000000

Table 7. Distribution of queue lengths at the pre-arrival epochs for the idle system $E_6/M^{5,10}/5$, $\rho = 0.1, 0.5, 0.9, \varepsilon = 10^{-20}$.

$\rho = 0.9$		n					Probabilities of k servers idle
		0	1	2	3	4	
	1	0.0045133	0.0062023	0.0077054	0.0090392	0.0102190	0.0376792
	2	0.0008471	0.0012652	0.0017914	0.0024039	0.0030832	0.0093908
$I(k)$	3	0.0000996	0.0001656	0.0002602	0.0003883	0.0005533	0.0014670
	4	0.0000058	0.0000116	0.0000210	0.0000351	0.0000554	0.0001289
	5	0.0000001	30.0000003	0.0000007	0.0000013	0.0000023	0.0000047
# in queue		0.0054659	0.0076450	0.0097787	0.0118678	0.0139132	SUM: 0.0486706
$\rho = 0.5$		n					Probabilities of k servers idle
		0	1	2	3	4	
	1	0.0476528	0.0526108	0.0551242	0.0557986	0.0551149	0.2663013
	2	0.0231853	0.0281785	0.0331282	0.0377295	0.0417906	0.1640121
$I(k)$	3	0.0064926	0.0089895	0.0118683	0.0150732	0.0185253	0.0609489
	4	0.0008326	0.0014196	0.0021965	0.0031834	0.0043942	0.0120263
	5	0.0000258	0.0000722	0.0001465	0.0002567	0.0004115	0.0009126
# in queue		0.0781891	0.0912706	0.1024636	0.1120414	0.1202365	SUM: 0.5042011
$\rho = 0.1$		n					Probabilities of k servers idle
		0	1	2	3	4	
	1	0.0005343	0.0002563	0.0001226	0.0000585	0.0000279	0.0009995
	2	0.009807	0.0057356	0.0033333	0.0019273	0.0011097	0.0219129
$I(k)$	3	0.0620315	0.0455535	0.0329312	0.0235189	0.0166373	0.1806724
	4	0.1067216	0.1061743	0.1006805	0.0923661	0.0827349	0.4886774
	5	0.0208943	0.0422757	0.0629306	0.0821285	0.0994901	0.0009996
# in queue		0.1999887	0.1999954	0.1999982	0.1999993	0.1999999	SUM: 0.9999815

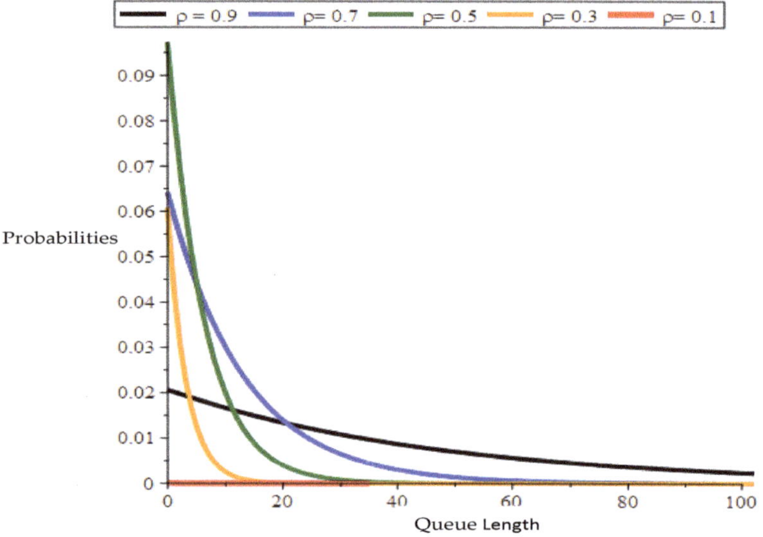

Figure 1. Comparison of performance measures of $D/M^{4,10}/5$ for busy servers, $\rho = 0.1, 0.3, 0.5, 0.7, 0.9, \varepsilon = 10^{-20}$.

Table 8. Comparison of performance measures of $E_6/M^{a,10}/5$, $D/M^{a,10}/5$, and $U/M^{a,10}/5$ for idle servers, $a = 1, 4, 7$, $\rho = 0.1, 0.5, 0.9$, $\varepsilon = 10^{-20}$.

	ρ	a	PI^-	L_Q^-	$SDLQ^-$	$E^-[I(k)]$	PI	L_Q	$SDLQ$	$E[I(k)]$	PI^+	L_Q^+	$SDLQ^+$	$E^+[I(k)]$
$E6/M^{a,10}/5$	0.1	1	0.5918	0.2683	0.7409	1.0643	0.4824	0.4082	0.8821	0.8093	0.7533	0.0000	0.0095	1.5576
		4	0.9998	1.5000	1.1181	3.8502	0.9997	1.5000	1.1181	3.7500	1.0000	1.3564	1.1081	4.4644
		7	1.0000	3.0000	2.0000	4.3430	1.0000	3.0000	2.0000	4.2857	1.0000	2.3522	1.9248	4.8083
	0.5	1	0.0205	5.8067	6.3981	0.0223	0.0107	6.2025	6.4242	0.0115	0.1047	1.3049	3.8825	0.1163
		4	0.3396	4.4994	5.6122	0.4972	0.3140	4.7317	5.7188	0.4523	0.6218	1.8572	3.1865	1.0733
		7	0.7796	3.7588	3.6688	1.6043	0.7670	3.8177	3.7469	1.5536	0.9205	3.0814	2.5287	2.4557
	0.9	1	0.0015	46.8450	47.4120	0.0016	0.0007	47.2590	47.4150	0.0008	0.0137	38.2720	46.5250	0.0145
		4	0.0285	45.6330	47.3290	0.0340	0.0249	46.0310	47.3480	0.0295	0.0991	37.4040	46.1810	0.1287
		7	0.1149	41.9250	46.7130	0.1636	0.1093	42.2760	46.7680	0.1543	0.2295	34.6680	44.9320	0.3837
$D/M^{a,10}/5$	0.1	1	0.6002	0.2327	0.6704	1.0461	0.4601	0.3998	0.8399	0.7354	0.7471	0.0000	0.0063	1.4800
		4	0.9999	1.5000	1.1180	3.8708	0.9998	1.5000	1.1800	3.7500	1.0000	1.3556	1.1077	4.4753
		7	1.0000	3.0000	2.0000	4.3548	1.0000	3.0000	2.0000	4.2857	1.0000	2.3415	1.9220	4.8142
	0.5	1	0.0182	5.6832	6.2591	0.0194	0.0076	6.1610	6.2870	0.0080	0.0924	1.2518	3.7652	0.1004
		4	0.3390	4.4119	5.4834	0.4914	0.3076	4.6909	5.6106	0.4374	0.6211	1.8256	3.0937	1.0579
		7	0.7842	3.7155	3.5716	1.6102	0.7690	3.7841	3.6631	1.5491	0.9229	3.0714	2.4851	2.4549
	0.9	1	0.0013	45.0490	46.6060	0.0014	0.0005	46.5470	46.6100	0.0005	0.0120	37.5580	45.7220	0.0125
		4	0.0280	44.8720	46.5260	0.0331	0.0237	45.3500	46.5470	0.0278	0.0983	36.7210	45.3830	0.1261
		7	0.1151	41.2030	45.9100	0.1630	0.1083	41.6240	45.9770	0.1518	0.2302	34.0240	44.1370	0.3819
$U/M^{a,10}/5$	0.1	1	0.6000	0.2338	0.6726	1.0468	0.4608	0.4001	0.8412	0.7378	0.7473	0.0000	0.0064	1.4825
		4	0.9999	1.5000	1.1180	3.8702	0.9998	1.5000	1.1180	3.7500	1.0000	1.3557	1.1077	4.4799
		7	1.0000	3.0000	2.0000	4.3544	1.0000	3.0000	2.0000	4.2857	1.0000	2.3419	1.9221	4.8140
	0.5	1	0.0182	5.6870	6.2634	0.0195	0.0076	6.1623	6.2913	0.0081	0.0928	1.2535	3.7689	0.1001
		4	0.3390	4.4117	5.4874	0.4916	0.3079	4.6921	5.6140	0.4378	0.6211	1.8266	3.0966	1.0584
		7	0.7840	3.7168	3.5746	1.6100	0.7690	3.7852	3.6657	1.5492	0.9229	3.0717	2.4865	2.4549
	0.9	1	0.0013	46.0740	46.6320	0.0014	0.0005	46.5690	46.6350	0.0005	0.0120	37.5810	45.7480	0.0125
		4	0.0280	44.8960	46.5500	0.0332	0.0238	45.3720	46.5720	0.0279	0.0983	36.7420	45.4080	0.1262
		7	0.1151	41.2260	45.9350	0.1631	0.1083	41.6440	46.0020	0.1519	0.2302	34.0450	44.1620	0.3819

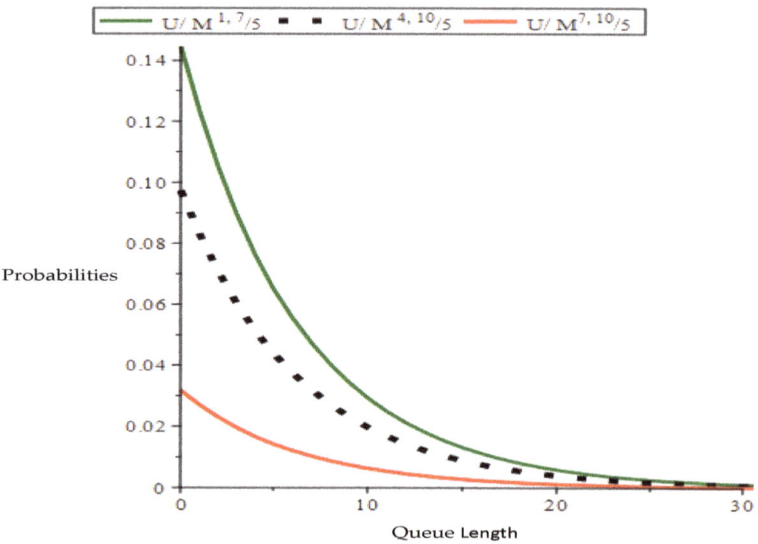

Figure 2. Comparison of performance measures of $U/M^{a,10}/5$ for busy servers, a = 1, 4, 7, $\rho = 0.5$, $\varepsilon = 10^{-20}$.

8. Conclusions

The queue $GI/M^{a,b}/c$ was successfully investigated by using the two-dimensional embedded Markov chain. Simple and exact analyses to determine queue-length distributions are presented. An algorithm was derived for the analysis of the steady state behaviour of the system. Our recursive solution approach is not only very efficient, but also accurate by providing the exact queue-length probabilities at p.a.e. In a similar manner, we studied the queue-length distribution at r.e. and derived closed-form formulae in terms of the root w for evaluating the exact queue-length probabilities at r.e. We also obtained the probabilities of p.d.e. through the relations between r.e. and p.d.e. The results for this system were provided numerically by considering three inter-arrival time distributions—Erlang, deterministic, and uniform. The work on higher order moments and other distributions can be conducted similarly.

There are two special features in this work. The first is the effort to express the important results in closed form; the second is the development of the methodology and algorithms to efficiently derive accurate results. The models under consideration were validated by using MAPLE to obtain numerical results with sufficient accuracy and trivial computational costs.

Author Contributions: The results in this paper are based on J.G.'s Ph.D. thesis Chapter 3. Conceptualization, methodology, writing—review and editing, funding acquisition, J.G. and M.C.; software, validation, formal analysis, writing—original draft preparation, J.G.; resources, supervision, M.C. All authors have read and agreed to the published version of the manuscript.

Funding: This research was supported by the Royal Military College of Canada Professional Development Allocation.

Institutional Review Board Statement: Not applicable.

Informed Consent Statement: Not applicable.

Data Availability Statement: Not applicable.

Conflicts of Interest: The authors declare no conflict of interest.

Appendix A. Algorithm for Calculating p.a.e. Probabilities

The method for determining the complete solution to the stationary queue-length probabilities at p.a.e. for the model $GI/M^{a,b}/c$ is described in the following steps:

1. Find the unique real root w inside the unit circle of Equation (14).
2. $X(B,n) = w^n, n \geq 0$. Let $k = 1$.
3. Calculate $X(I(1), a-1)$ by using Equation (20).
4. Calculate $J(k)$ by using Equation (19).
5. Calculate $X(I(k), a-2), \cdots, X(I(k), 0)$ recursively by using Equation (21).
6. Substitute $X(I(k), a-1)$ and $X(I(k), 0)$ into Equation (22) to find $X(I(k+1), a-1)$. Let $k = k+1$.
7. Repeat step 4 to step 6, and solve for the rest of the idle server probabilities.
8. Finally, find the normalized p.a.e. vector using $\mathbf{P}^- = \frac{X}{C_N}$.

Appendix B. Proof of Equation (25)

$$D_R(z) = \sum_{l=0}^{\infty} (l|c)_R z^l = \frac{\rho b}{1-z}[1 - \bar{a}(c\mu(1-z))].$$

Proof.

$$\sum_{l=0}^{\infty} (l|c)_R z^l = \sum_{l=0}^{\infty} z^l \int_0^{\infty} \frac{e^{-c\mu t}(c\mu t)^l}{l!} dR(t)$$

$$= \int_0^{\infty} e^{-c\mu t} \sum_{l=0}^{\infty} \frac{(c\mu t z)^l}{l!} dR(t)$$

$$= \int_0^{\infty} e^{-c\mu t} e^{c\mu t z} dR(t)$$

$$= \int_0^{\infty} e^{-c\mu(1-z)t} \lambda(1 - A(t)) dt$$

$$= \lambda \underbrace{\int_0^{\infty} e^{-c\mu t(1-z)} dt}_{=1/c\mu(1-z)} - \lambda \int_0^{\infty} e^{-c\mu(1-z)t} A(t) dt$$

$$= \frac{\rho b}{1-z} + \frac{\rho b}{1-z} \int_0^{\infty} A(t) de^{-c\mu(1-z)t}$$

$$= \frac{\rho b}{1-z} \Big(1 - \underbrace{\int_0^{\infty} e^{-c\mu(1-z)t} dA(t)}_{=\bar{a}(c\mu(1-z))}\Big). \qquad \text{(using Equation (11))}$$

$\square$

References

1. Shyu, K. On the queueing processes in the system GI/M/n with bulk service. *Acta. Math. Sin.* **1960**, *10*, 182–189.
2. Gross, D.; Shortle, J.; Thompson, J.; Harris, C. *Fundamentals of Queueing Theory*; John Wiley and Sons: Hoboken, NJ, USA, 2008.
3. Neuts, M. A general class of bulk queues with Poisson input. *Ann. Math. Stat.* **1967**, *38*, 759–770. [CrossRef]
4. Easton, G.; Chaudhry, M. The queueing system $E_k/M^{a,b}/1$ and its numerical analysis. *Comput. Oper. Res.* **1982**, *9*, 197–205. [CrossRef]
5. Chaudhry, M.; Madill, B. Probabilities and some measures of efficiency in the queueing system $GI/M^{a,b}/1$. *Selecta Statistica Canadiana* **1987**, *7*, 53–75.
6. Neuts, M. *Matrix-Geometric Solution in Stochastic Models—An Algorithmic Approach*; John Hopkins University Press: Baltimore, MD, USA, 1981.
7. Sasikala, S.; Indhira, K. Bulk service queueing models—A survey. *Int. J. Pure Appl. Math.* **2016**, *106*, 43–56.
8. Medhi, J. Further results on waiting time distribution in Poisson queue under a general bulk wervice rule. *Cahiers du C.E.R.O.* **1979**, *21*, 183–189.
9. Sim, S.; Templeton, J. Steady state results for the $M/M^{a,b}/c$ batch-service system. *Eur. J. Oper. Res.* **1985**, *21*, 260–267. [CrossRef]

10. Sim, S. On Multi-Vehicle Transportation Systems with Queue-Dependent Dispatching Policies. Ph.D. Thesis, University of Toronto, Toronto, ON, Canada, 1982.
11. Adan, I.; Resing, J.C. Multi-server batch-service systems. *Stat. Neerl.* **2000**, *54*, 202–220. [CrossRef]
12. Goswami, V.; Samanta, S.; Vijaya Laximi, P.; Gupta, U. Analyzing a multiserver bulk-service finite-buffer queue. *Appl. Math. Model.* **2008**, *32*, 1797–1812. [CrossRef]
13. Shyu, K. The waiting time distribution for the queueing processes in the system GI/M/n with bulk service. *Acta. Math. Sin.* **1964**, *14*, 796–808.
14. Chaudhry, M.; Templeton, J. *A First Course in Bulk Queues*; John Wiley & Sons, Inc.: New York, NY, USA, 1983.

mathematics

Article

Processing Large Outliers in Arrays of Observations

Gurami Tsitsiashvili

Institute for Applied Mathematics, Far Eastern Branch of Russian Academy of Sciences, 690041 Vladivostok, Russia; guram@iam.dvo.ru; Tel.: +7-914-693-2749

Abstract: The interest in large or extreme outliers in arrays of empirical information is caused by the wishes of users (with whom the author worked): specialists in medical and zoo geography, mining, the application of meteorology in fishing tasks, etc. The following motives are important for these specialists: the substantial significance of large emissions, the fear of errors in the study of large emissions by standard and previously used methods, the speed of information processing and the ease of interpretation of the results obtained. To meet these requirements, interval pattern recognition algorithms and the accompanying auxiliary computational procedures have been developed. These algorithms were designed for specific samples provided by the users (short samples, the presence of rare events in them or difficulties in the construction of interpretation scenarios). They have the common property that the original optimization procedures are built for them or well-known optimization procedures are used. This paper presents a series of results on processing observations by allocating large outliers as in a time series in planar and spatial observations. The algorithms presented in this paper differ in speed and sufficient validity in terms of the specially selected indicators. The proposed algorithms were previously tested on specific measurements and were accompanied by meaningful interpretations. According to the author, this paper is more applied than theoretical. However, to work with the proposed material, it is required to use a more diverse mathematical tool kit than the one that is traditionally used in the listed applications.

Keywords: large outliers; arrays of observations; complex systems; digraphs

MSC: 93B07; 06A06

Citation: Tsitsiashvili, G. Processing Large Outliers in Arrays of Observations. *Mathematics* **2022**, *10*, 3399. https://doi.org/10.3390/math10183399

Academic Editor: Andrzej Sokołowski

Received: 18 August 2022
Accepted: 15 September 2022
Published: 19 September 2022

Publisher's Note: MDPI stays neutral with regard to jurisdictional claims in published maps and institutional affiliations.

1. Introduction

This paper is devoted to the analysis of large outliers in data samples in medical and zoo geography, mining, an application of meteorology in fishing tasks, etc. The closest to this problem in probability theory, mathematical statistics, queuing theory and insurance is the analysis of heavy-tailed distributions [1–7].

It should be noted that recently, this topic has attracted the attention of a large number of data processing specialists from the fields of mathematical statistics [8,9], statistical methods in medicine [10,11] and physiological studies [12], as well as in the analysis of industrial processes [13,14]. Moreover, along with the statistical methods in this area, it requires the development of new algorithms and the application of graph theory elements, particularly in the study of protein networks [15].

However, in those applications with which the author had to work, it was necessary to shift the emphasis from estimates of heavy tails to the large outliers in empirical information. Apparently, this is due to the fact that we have to work with short samples or in the presence of rare events. However, the main reason is that there are no well-established theoretical models in these areas of application, and we have to work with data within the framework of a phenomenological approach. This circumstance required the development of original heuristic algorithms that allowed obtaining information useful and interesting to users who submitted their empirical results to the author. The novelty and significance of the algorithms constructed by the author were confirmed during last 20 years by the joint results represented in [13,15–20]. Previously, for a long time, such tasks simply could not be solved by the author.

In the listed areas of application, the ability to consistently meet the user requirements plays a crucial role. The following motives are important for these specialists: the substantial significance of large emissions, the fear of errors in the study of large emissions by standard and previously used methods, the speed of information processing and the ease of interpretation of the results obtained. To meet these requirements, interval pattern recognition algorithms and the accompanying auxiliary computational procedures have been developed. These algorithms were designed for specific samples provided by the users. They have the common property of the original optimization procedures being built for them or well-known optimization procedures being used.

The emphasis on large outliers is due to the fact that their behavior usually obeys some asymptotic relations [21] and is therefore somewhat simplified. Such circumstances allow us to raise the question of increasing the reliability of the results of the processing arrays of observations and reducing the counting time. The latter plays an important role in the interdisciplinary interactions between domain specialists and mathematical programmers processing the arrays of observations. To carry out such work, it is advisable to identify the applied tasks in which such observation processing procedures may be implemented.

The considered samples of observations are defined by the number n of observations and the number m of their dimensions. The requirements of mathematical statistics [10] are such that it is desirable that the parameter n is large and the parameter m is small. However, in the arrays of observations with which we had to deal, the opposite situation was often observed, where the parameter n was small and the parameter m was large. For example, such a situation occurs in problems of medical geography [17] and in problems of meteorology and hydrology [18]. This circumstance forces one to look for sufficiently fast algorithms for processing short time series, and the accuracy of calculations determined in some way, on the contrary, increases with an increase in the parameter m.

At the same time, there are one-dimensional long time series ($m = 1$, and n is sufficiently large) in which not just rare but very rare events associated with large outliers are observed [13]. It is required to process these series in such a way that the length of the series and the number of large outliers in it do not create problems for either processing or interpretation of the results obtained.

Along with time series, which are not quite convenient for data processing, in various applications, there are large arrays of observations that require data compression and packaging and lead to extreme graph theory problems. These include disturbances in the rock according to the results of acoustic monitoring and the movement of animals in a territory. Despite the presence of well-known graph theory algorithms, special auxiliary algorithms, albeit simple, are designed well enough with the requirements of a particular subject area and are also required for processing such data.

This paper describes the methods of interval pattern recognition used in medical geography and meteorology recognition of rare outliers by a generalized indicator used in mining, studies of the vicinity of the extremes in the nodes of the square grid used in meteorology and hydrology and special classification methods used in the analysis of protein networks in zoo geography, mining and other subject areas.

2. Materials and Methods

The materials for constructing algorithms for processing empirical information are the following:

- Multidimensional short series of observations containing the main component and m accompanying components;
- Series of real observations equipped with Boolean variables indicating the presence or absence of critical events;
- An array of three-dimensional vectors characterizing the coordinates of sound sources;
- An array of one-dimensional characteristics of square lattice nodes;

- A scheme of the protein network in the form of a digraph;
- A map with a set of districts and a description of the presence or absence of borders between them.

The methods are as follows:

- The method of interval pattern recognition;
- The method for optimizing monotone piecewise constant functions;
- The method for converting a matrix of distances between points in three-dimensional space into an undirected graph;
- The method for difference approximation of first- and second-order partial derivatives for the functions of two variables;
- The method of sequentially allocating cyclic equivalence classes in a digraph and constructing a zero-one matrix of a partial order between these classes;
- The method of hierarchical classification of districts on a map with respect to the presence of common boundaries between them.

The following optimization problems are considered:

- When recognizing critical events from an array of one-dimensional observations, two optimization problems are considered. A connection between them is established, and it is shown how by reducing one task to another, the array of processed information may be significantly decreased.
- In determining the acoustic core, the connectivity component that contains the minimum number of vertices is selected from another connectivity component of a graph.
- An algorithm for approximating a level line of a smooth function, given at the nodes of a square lattice, in the form of an ellipse is constructed.
- In the hierarchical classification of districts on a map, for each district, the minimum number of borders, a crossing of which allows one to get out from this district to a common boundary, is determined.

All described methods are closely connected with the initial formulations of the applied problems and are adopted to real data processing. Moreover, in relation to each case, it is necessary to introduce some new element into the algorithm.

3. Interval Pattern Recognition Method and Related Algorithms

This section discusses the interval pattern recognition algorithm, which has found its application in the processing of time series in the problems of medical geography [17], as well as in meteorology, hydrology [18] and fishing [22,23].

3.1. Interval Pattern Recognition Method

Suppose that an array of observations is represented by a set of vectors with dimensions $m + 1$: $X = \{(x_{01}, x_{11}, \ldots, x_{m1}), \ldots, (x_{n0}, x_{n1}, \ldots, x_{nm})\}$. Here, the components of vectors $x_{01}, \ldots, x_{0m}$ characterize the main features, and all other components of these vectors are related features. Let us say the element $(x_{k0}, x_{k1}, \ldots x_{km})$ corresponds to a larger outlier in the sample if the inequality $x_{k0} \geq x_0$ is satisfied at some critical level x_0 (selected by an expert) of the zero component value in the vector. Then, in the initial sample X, a set of elements with numbers $1 \leq k_1, \ldots, k_s \leq n$ is determined, for which the inequality $x_{k_j0} \geq x_0$, $1 \leq j \leq s$ is satisfied. All these elements are perceived as large outliers. We first calculate

$$x_i^+ = \max_{1 \leq j \leq s} x_{k_j i}, \ x_i^- = \min_{1 \leq j \leq s} x_{k_j i}, \tag{1}$$

Then, a decisive rule is constructed according to which the sample element $(x_{k0}, x_{k1}, \ldots x_{km})$ is a large outlier if the following inequalities are met:

$$x_i^- \leq x_{ki} \leq x_i^+, \ 1 \leq i \leq m. \tag{2}$$

This decisive rule is defined as interval pattern recognition. Here, the image is understood as a large outlier determined by the value of the zero component of the sample element, and the decisive rule (2) is determined by the belonging of the components of the vector $(x_{k0}, x_{k1}, \ldots x_{km})$ to the segments $[x_i^-, x_i^+]$, $1 \leq i \leq m$.

Let us now list the main properties of interval pattern recognition. For this, we denote S as the number of sample elements that are perceived by this decisive interval recognition rule as large outliers:

- All sample elements that are large outliers are perceived by interval recognition as large outliers. Therefore, the $S \geq s$ inequality is fulfilled. Then, the quality of interval recognition may be chosen by the ratio $s/S \leq 1$.
- With an increase in the number m of associated features, the recognition quality of s/S increases and, for some samples of observations, may even approach unity.
- The number of arithmetic operations for the interval recognition procedure is proportional to the product nm and therefore depends linearly on the number n of sample elements X and on the number m of accompanying features.
- The solution of this problem in its initial version was tested with respect to $s/S_{,,}$ characterizing the quality of recognition for a given sample. Here, it is possible to increase the value 0.6 obtained by standard methods to 0.7 or more with an increase in the number m.

3.2. Investigation of the Extremum of a Function in the Nodes of a Square Lattice

The most important element of a structure of the pressure field at an altitude of 5 km above the Far East is a stable and extensive depression. The coordinates of this depression (which are usually associated with a square lattice node) and the pressure value H_{500} at an altitude of 5 km determine the nature of atmospheric circulation and significantly affect the weather [19]. This also includes observations represented by a finite number of points located at the nodes of a square lattice and characterizing a certain meteorological system. It is known from observations that the extremes of H_{500} at the nodes of such a grid largely determine the functioning of the meteorological system. If we assume that H_{500} is described by a smooth function defined on a rectangle and having a minimum at the lattice node, then by decomposing this function into a Taylor series and assuming the lattice step is small enough, we may approximate the level lines of this function with ellipses [19]. In turn, the direction of the major axis of the ellipse and its relation to the minor axis allow us to make meteorological forecasts concerning the behavior of anticyclones in the vicinity of the minimum point.

Suppose that the function $f(x, y)$, specifying H_{500}, is continuously differentiable twice in the domain $D = \{0 \leq x \leq Nh, \ 0 \leq y \leq Mh\}$, and at the point $(kh, lh), 0 < k < N$, $0 < l < m$, its first differential is zero, and its second differential $A(x - kh)^2 + B(y - lh)^2 + 2C(x - kh)(y - lh)$ is a positive definite quadratic form ($A = f_{x,x}(kh, lh)$, $B = f_{y,y}(kh, lh)$, $C = f_{x,y}(kh, lh)$). Then, the point (kh, lh) is the point of the local minimum of the functions f, and therefore, by virtue of the Sylvester criterion, the inequalities $A + B > 0$, $AB > C^2$ are fulfilled. The lines of the level of the function f in the vicinity of the point (kh, lh) are approximately ellipses. The angle of inclination of the major axis and the compression ratio of these ellipses determine the nature of the atmospheric circulation.

We denote $a = A + o(h)$, $b = B + o(h)$ and $c = C + o(h)$ as the finite difference approximations of the partial derivatives A, B and C. We approximate the function f by the function $\widehat{f}$ up to $o(h^2)$ in variables $X = \dfrac{x - kh}{h}$ and $Y = \dfrac{y - lh}{h}$:

$$\widehat{f}(x, y) = f(kh, lh) + \frac{1}{2}(aX^2 + bY^2 + 2cXY), \ a + b > 0, \ ab > c^2.$$

Therefore, for small h values, the quadratic form $aX^2 + bY^2 + 2kxy$ is also positively definite.

We reduce this form to a diagonal form by constructing a matrix $\mathcal{A} = \begin{pmatrix} a & c \\ c & b \end{pmatrix}$ and writing out the characteristic equation $(a - \lambda)(b - \lambda) - c^2 = 0$, whose roots

$$\lambda_\pm = \frac{a+b}{2} \pm \sqrt{\left(\frac{a+b}{2}\right)^2 - ab + c^2} > 0,$$

are the eigenvalues of the matrix $\mathcal{A}$.

In the coordinate system (u_+, u_-) with an orthonormal basis $\overrightarrow{n}_+$, $\overrightarrow{n}_-$ from the eigenvectors of matrix $\mathcal{A}$, the quadratic form $aX^2 + bY^2 + 2cXY$ is represented by the sum of squares $\lambda_+ u_+^2 + \lambda_- u_-^2$ with level lines in the form of ellipses $\lambda_+ u_+^2 + \lambda_- u_-^2 = const > 0$, having a compression ratio $k = \sqrt{\lambda_+/\lambda_-}$. The slopes of the major axis of the ellipses were found, and the compression ratio at the H_{500} level line allowed meteorologists to build a physical reconstruction of various processes occurring in the atmosphere. The lines of the level of the analyzed function H_{500}, constructed in the form of ellipses, were rechecked during the construction of a physical meteorological forecast in [19].

4. Recognition of Rare Outliers and Related Algorithms

Another type of observation may be time series in which $m = 1$ and the length of the series n is quite large, being to the order of several hundred. Such observations characterize important and therefore rare events in the system. These include the already described collapses in mine workings. The miners proposed to characterize the state of the system at some point in time by a generalized one-dimensional indicator ρ and a Boolean variable characterizing the presence or absence of a collapse in the system. The task is to recognize presence or absence of the collapse in the presented one-dimensional series of observations. An algorithm is proposed for constructing a recognition procedure for the presence or absence of the collapse, in which the amount of calculations is determined only by the number of important events N being much smaller than n. This algorithm is based on maximizing the frequency of correct recognition of the presence or absence of an event from the critical value ρ_*, determining the recognition result using the inequality $\rho \leq \rho_*$.

Let us now turn to the consideration of long series of observations in which the number of large emissions is small (i.e., n is much larger than one, and N/n is much smaller than one). Such observations include, in particular, collapses in mine workings. There is a class of applied problems in which a certain generalized indicator is selected as a concomitant feature, formed by specialists of this subject area based on the results of numerous observations, such as mining specialists based on the results of acoustic monitoring of the rock strata [13,24,25].

4.1. Recognition of Rare Outliers by a Generalized Indicator

In this subsection, we assume that the initial sample is formed as follows. All generalized indicators form a sequence $\{x_{11}, \ldots, x_{n1}\}$, and the numbers $k_1, \ldots, k_s$ of the sample elements characterizing large outliers are given. It is required to build a recognition rule for determining emissions by this generalized (single) indicator. Let us place the sequence $\{x_{11}, \ldots, x_{n1}\}$ on the real line and mark it with crosses with the numbers $k_1, \ldots, k_s$ (see Figure 1). We are looking for a number x_* defining the following decisive rule: if $x_k \geq x_*$, then the sample element with the number k refers to large outliers. If $x_k < x_*$, then the sample element with the number k is not recognized as a large outlier.

Figure 1. Representation of a training sample on a straight line by a set of characters $\times$, $\bullet$.

For each number c, we compare the frequency $\rho_\times(c)$ of correctly attributing a sample element to large outliers and the frequency $\rho_\bullet(c)$ of not correctly attributing a sample element to large outliers. The value ρ_* is introduced using an expert method, and it is required that the solution corresponding to it satisfies the inequality $\rho_\times(c) \geq \rho_*$. Among all $c : \rho_\times(c) \geq \rho_*$, it is required to find one value that maximizes $\rho_\bullet(c)$. Here, $\rho_\times(c)$ characterizes the security of the decision being made, and $\rho_\bullet(c)$ characterizes its cost-effectiveness.

Since the function $\rho_\times(c)$ is stepwise and monotonically non-increasing by the argument c, being continuous to the left, and the function $\rho_\bullet(c)$ is stepwise and monotonically non-decreasing, being continuous on the right (see Figures 1–3), then this problem has many solutions that can be represented by some segment. In turn, the task of determining the maximum value of x_* at which $\rho_\times(c) \geq \rho_*$ has a unique solution, which is the right end of the segment specified above. It is natural, for security reasons, to determine the right end of the segment, which is the solution to the maximization problem $\rho_\bullet(c)$, under the condition $\rho_\times(c) \geq \rho_*$. Due to the specified property of this solution, it is sufficient to solve the problem for the maximum of the function $\rho_\times(c)$ under the condition $\rho_\times(c) \geq \rho_*$.

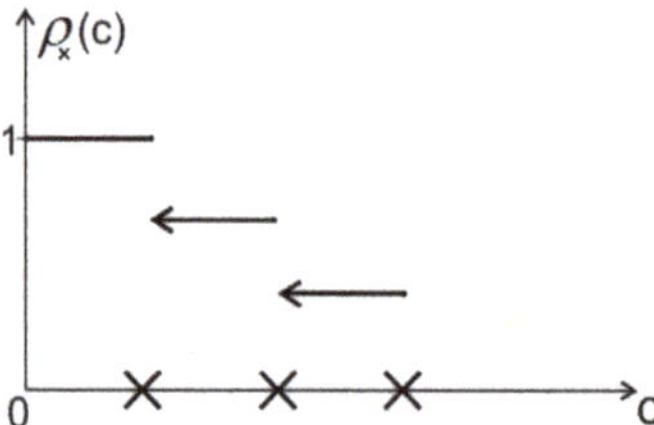

Figure 2. Type of function $\rho_\times(c)$.

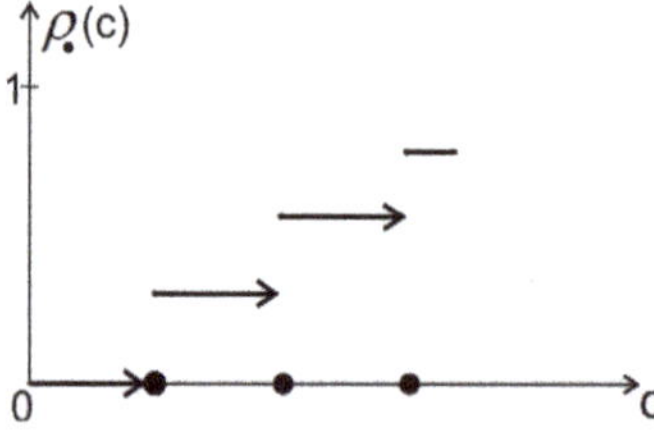

Figure 3. Type of function $\rho_\bullet(c)$.

The resulting solution to the problem of recognizing large outliers by sampling $\{x_{11}, \ldots, x_{n1}\}$ and numbers $k_1, \ldots, k_s$ requires only knowledge of the sequence $x_{k_1 1}, \ldots x_{k_s 1}$, which significantly reduces the amount of calculations, since s/n is much smaller than one.

Using the method of recognizing a large outburst (exceeding the generalized indicator of the critical level), the results were obtained for predicting collapses in the mine, which were confirmed by specialists in mining. Moreover, the frequency of correct recognition of a critical event (a collapse in a mining operation) constructed in solving this problem characterizes the safety factor of mining operations, and the frequency of correct recognition of an absence of a critical event characterizes the cost-effectiveness factor of mining operations. Therefore, when solving this problem, safety restrictions were first introduced, and under these restrictions, the efficiency indicator was optimized. The solution of the concrete problem considered in [13] was verified by comparing the optimization result c obtained by the author and the result independently obtained by mining specialists (which practically coincided). It was very important for the mining specialists to independently verify their own rather cumbersome calculations.

4.2. Clusters of Points in Space

When implementing an acoustic monitoring system, it becomes necessary to determine acoustically active zones and, on this basis, predict dangerous collapses in mining according to the generalized indicator introduced by mining specialists [24,25]. In the previous subsection, to construct a generalized indicator, it was necessary to determine the acoustically active zone from a set of n points in three-dimensional space determined during acoustic monitoring of cod sources in the rock column [13].

In fact, we are talking about constructing a model of an acoustically active zone based on the available observations and an algorithm for determining it. This procedure is based on information about the matrix $||r_{ij}||_{i,j=1}^{n}$ of pairwise distances between the points detected during acoustic sounding and the critical distance r between them, as set by experts.

For the first step, the matrix $||r_{ij}||_{i,j=1}^{n}$ is converted to a zero-one matrix $||I(r_{ij} < r)||_{i,j=1}^{n}$.

In the second step, the constructed zero-one matrix is further considered as the adjacency matrix of an undirected graph, whose edges between the vertices i and j exist only under the condition $r_{ij} < r$.

In the third step, using the well-known methods of graph theory, in the set of $1, \ldots, n$ vertices of the graph G, a set of connectivity components is determined, among which the one with the maximum number of vertices is selected. This set of vertices is defined as an acoustically active zone (several zones are also possible).

In the forth step, the classification procedure is accelerated in the following way. Initially, the point 1 is taken, which is denoted by the first class. Let the vertex classes $I_1, \ldots, I_p$ from the set $\{1, \ldots, k\}$ be allocated in step k, and the point $k + 1$ is connected by the edges to some of these classes. Then, a new class is formed from them and the point $k + 1$, and the classes that are not included in this new class remain the same, together forming a set of classes in step $k + 1$. In such an algorithm, information previously used is not lost at each step of the algorithm. The most significant step is the last step of this algorithm, in which it is proposed to preserve the classification of the connectivity components of the graph and not leave only one applicant for the formation of the final connectivity component.

The selection of clusters of points in the three-dimensional space detected during acoustic monitoring allows us to build generalized indicators by which critical events (collapses) in a mine are predicted. The solution to the problem considered in [13] was verified visually by mining specialists, who were interested in convenient computer algorithms for defining acoustically active zones.

5. Special Classification Algorithms

Classification algorithms allow us to identify some extreme modes in a complex system. In particular, with the help of classification algorithms, it is possible to determine the acoustically active zones. Of particular interest are hierarchical classification algorithms that identify objects, namely those that most influence the behavior of a complex system or objects that play the role of hubs through which numerous connections between elements pass. This section of the work is devoted to these issues.

5.1. Hierarchical Classification of Graph Vertices

This problem arose when analyzing a protein network presented by a complete digraph containing n vertices [26]. The vertices of such digraphs are proteins and the directed edges of the connection between them. The procedure of hierarchical classification in such a digraph is in some sense equivalent to the isolation of clots (aggregates of proteins close to each other).

Using Floyd's algorithm, we construct a matrix $||c_{ij}||_{i,j=1}^{n}$ of the lengths of the minimal paths between the vertices of the original digraph. We transform the matrix $||c_{ij}||_{i,j=1}^{n}$ into a symmetric matrix $||r_{ij}||_{i,j=1}^{n}$, $r_{ij} = c_{ij} + c_{ji}$. Thus, r_{ij} is the minimum length of a cycle connecting the vertices i and j. It is obvious that the minimum length of a cycle passing through a pair of vertices can be considered the distance between them, since it is nonnegative and satisfies the triangle inequality.

Let us construct a finite, monotonically increasing sequence of $R = \{r_1 < r_2 < \ldots < r_m\}$ nonzero elements of this matrix. Having chosen some critical level r, we transform the matrix $||r_{ij}||_{i,j=1}^{n}$ into a zero-one matrix $||I(r_{ij} \le r)||_{i,j=1}^{n}$. Now, let us construct a graph G_r, whose edges connect the vertices i and j provided that $r_{ij} \leq r$. Then, in the undirected graph G_r, the connectivity components may be distinguished using the classification algorithms described above. The parameter r may be selected in different ways, such as by assuming $r = r_1, \ldots, r_m$. In this case, with $1 \leq r_p < r_q \leq r_m$, the class defined with $r = r_q$ necessarily enters some class defined with $r = r_p$. Thus, the hierarchical classification of the set of vertices $1, \ldots, n$ is determined. However, the increasing sequence of values of the critical level r may be reduced at the choice of the users.

5.2. Allocation of Cyclic Equivalence Classes in a Digraph

The problem considered above requires for its formulation the allocation of cyclic equivalence classes (clusters) in the digraph. The cyclic equivalence relation between a pair of digraph vertices assumes the existence of a cycle containing this pair of vertices. Then, a partial order relation may be introduced between the cyclic equivalence classes in the digraph. There are different algorithms to define the cyclic equivalence classes and matrix of their partial order (see, for example, [27,28]).

In order to construct a sequential algorithm for solving this problem, it is required at each step to establish a partial order relation between the classes of cyclic equivalence. It is not enough to just allocate cyclic equivalence classes. It is also required to determine the zero-one matrix of the partial order relationship of clusters (a presence of a path from one cluster to another).

To accomplish this, at step 1, the vertex 1 is taken, and a cluster and a one-by-one unit matrix are formed from it. Let the clusters and the matrix of partial order relations between them be constructed at step $t - 1$. We take the element t and select the following sets of clusters: B_1, B_2 and B. The set B_1 contains clusters, each of which has a path from the vertex t, and the set B_2 contains clusters from which there are paths to the vertex t. All other clusters fall into the set B, and from them, there can be paths only to the clusters of the set B_1, and paths can exist in them only from the set B_2. Then, at step t, a new cluster $[t]$ is built, consisting of the vertex t and the clusters of the set $B_1 \cap B_2$. The matrix of a partial order at step t is defined by rectangular sub matrices 0 consisting of only zeros, rectangular sub matrices 1 consisting of only ones and rectangular sub matrices repeating the corresponding submatrices of the matrix a at step $t - 1$ (see Table 1).

Table 1. Algorithm of transition from step $t - 1$ to step t for a matrix of partial order a.

Matrice Partial Order	Clusters Set A_1	Clusters Set $[t]$	Clusters Set A_2	Clusters Set B
set A_1 clusters	repeating step $(t - 1)$	0		0
set $[t]$ clusters		1		
set A_2 clusters			repeating step $(t - 1)$	
set B clusters	repeating step $(t - 1)$	0	repeating step $(t - 1)$	

This method has been applied to the analysis of the thermal stability of some protein networks [16], and so far, requests have been received from various applied biological journals for the continuation of this topic.

5.3. Definition of Central Hub Areas on the Map

Another type of such observations may be maps divided into some areas and used to highlight areas associated with animal movements [29]. Let us assume that there is some bounded, connected territory with a set of U_0 singled-out, single-connection regions (administrative districts or hunting farms) on it. This territory is defined by a finite set of bounded regions on the plane. Everywhere else, without limiting generality, we assume that the boundaries between the regions are polylines. Our task is to compress information about this map in order to use it further for studying the movement of rare animals in this area by traces of these animals found in these areas.

According to this division, it is necessary to build a hierarchical classification of internal (not touching the border of the map) districts in relation to their neighborhood. Such a hierarchical classification assumes the allocation on the map of a sequence of sets of districts U_k, $U_{k+1} \subseteq U_k$, $k \geq 1$ so that each district in the set U_{k+1} adjoins only the districts from the set U_k. It is shown that such a sequence is finite, and in real observations, the number of vertices at the end of the algorithm is usually significantly less than at the beginning. Thus, the final vertices allow us to compress the original information about the map.

This compression of map information is based on the "neighborhood" relationship between the specified areas. For this purpose, a map with the areas highlighted on it is represented as a planar graph, the faces of which are the regions, and the edges are the sections of the border between two neighboring regions.

This procedure can be continued in a recurrent way:

$$U_{k+1} = \{A \in U_k : S(A) \subseteq U_k\}, \, k \geq 1 \tag{3}$$

This can continue up to some step n, at which point one of two equalities is fulfilled: $U_{n+1} = U_n$ or $U_{n+1} = \varnothing$. Here, for $A \in U_0$, we define $S(A)$ as a set of regions bordering it.

The equality $U_{n+1} = U_n$ means that all regions of the set U_n border only on the regions of this set. However, due to the condition of the limitation of all areas of the map, the finiteness of the number of these areas and the presence of only polylines as boundaries, this condition cannot be fulfilled. In addition, since at each step k the strict inclusion of $U_{k+1} \subset U_k$ is performed, then the number of regions $N(U_k)$ in the set U_k satisfies the inequality $N(U_{k+1}) < N(U_k)$. This implies the inequality $n < N(U_0) < \infty$. Therefore, the algorithm in Equation (3) may be implemented in a finite number of steps n. In the second case, when $U_{n+1} = \varnothing$, we have $N(U_{n+1}) = 0$, so no area from the set U_n may be completely surrounded by areas from the same set. This algorithm requires knowledge of the set of all inner regions U_1 and the sets $\{S(A) : A \in U_1\}$ of all regions bordering the inner regions (of the first kind). Thus, the implementation of the algorithm in Equaiton (3) is working with lists of the area numbers and not with their view on the plane, which greatly simplifies its implementation.

Denote $V_k = U_k \setminus U_{k+1}$, $1 \leq k < n$, $V_n = U_n$, and then the equalities are valid ($U_k = \bigcup_{j=k}^{n} V_j$, $1 \leq k \leq n$), and any vertex of the set V_k is connected by an edge to some vertex of the set V_{k-1} where there are no edges connecting this vertex to the vertices of the sets V_j, $j < k - 1$. Indeed, if the vertex is $v \in V_k$, then the inclusion of $v \in U_k$ is performed. However, a complete encirclement of a vertex v by vertices from the set U_k is impossible, because in this case, $v \in U_{k+1}$ means $v \in V_j$ for some $j \leq k - 1$. Therefore, there is an edge connecting the vertex v with the set of vertices U_{k-1}. However, an edge connecting the vertex v to the set U_{k-2} is also impossible, because the vertex v is completely surrounded by the vertices of the set V_{k-1}. Finally, the vertex $v \in V_k$ may be connected with some vertices of this set also. Therefore, each region of the set $U_n = V_n$ may be considered some center on the map. Then, the set V_{n-1} consists of the areas bordering it and completely surrounding it, called its margin or periphery of the first kind. By attaching to the periphery of the first kind, with the regions bordering on the regions from this periphery, it is possible to build a periphery of the second kind, and so on. It follows from this construction that the minimum number of boundaries that the path from the vertex $v \in V_k$ to the total boundary of all districts crossed is equal to k, where $k = 1, \dots, n$. The proposed algorithm was tested during the analysis of traces of the Amur tiger in the territory of Primorsky Krai with the help of ecologists and aroused their serious interest.

6. Discussions

What all the algorithms for processing large outliers given in this paper have in common is the fact that the algorithms themselves are fairly standard, but when applying them to individual samples, it is necessary to select the correct combination of these algorithms. It is this combination that ensures the novelty of the results obtained. For example, when processing data on acoustic monitoring of the rock strata for an algorithm for predicting critical events (collapses), an algorithm for identifying acoustically active zones was required. In turn, when analyzing critical events in the climate system, it is necessary not only to highlight the moments of occurrence of these events but also their spatial localization and behavior in its vicinity.

In the practical application of the proposed algorithms, their computational complexity and computational speed play an important role. In some cases, for example, when processing data on animal movements over a certain territory, excessive requirements for data processing algorithms may encounter excessive computational complexity. This led to the construction and use of hierarchical classification algorithms, which at the top level of the hierarchy identify some central parts of the study area.

The final results of the proposed algorithms for processing observations are evaluated by experts from the subject area. Therefore, all elements of the proposed algorithms should be understood by these experts and allow them to be checked. Moreover, the proposed algorithms should be convenient to assist experts in constructing various scenarios of the behavior of the analyzed system. It should be noted that the results of processing large outliers tend to be some estimates that require estimates of their errors and the impact of the inaccuracies of the observations of them.

The experience of working with algorithms for processing large outliers shows that all the elements included in them should be selected as carefully as possible in order to ensure high quality and demand among specialists in the subject areas. It is also necessary to combine the proposed algorithms for processing large outliers with classical probabilistic models. For example, when processing data on animal tracks in a certain territory, it is convenient to use an inhomogeneous Poisson flow of points [30] as a model of animal tracks. Now, it is difficult to predict what new algorithms and models will have to be built to solve the problems discussed in the work. These tasks come from users and require additional mathematical processing, but it is already clear that various optimization procedures should play an important role in them.

When identifying flashes in a time series, some difficulties arise that require a set of different methods to overcome. For example, there are known time series of pink salmon yields, in which the harvest is small in even years and large in odd years. To analyze this phenomenon, it is necessary to distinguish stable cycles of a length of two in the Ricker model. These cycles appear when the growth coefficient of the model belongs to a certain interval. However, the noted phenomenon occurs only at the right end of the interval, and this can be detected only after additional and more detailed calculations.

7. Conclusions

This article presents an algorithm for constructing an interval pattern recognition procedure. The properties of this algorithm were investigated, and it was shown that with an increase in the dimension of observations, the recognition quality improves:

- An algorithm for recognizing a critical event from a one-dimensional series of observations was constructed by analyzing the (small) part of the series containing only critical events.
- An algorithm for determining the acoustically active zone by the coordinates of the sound source points was constructed. This algorithm is based on the transformation of an array of coordinates of sound source points into an undirected graph and the allocation of connectivity components in it.
- A sequential algorithm for determining cyclic equivalence classes and partial order relations between these classes in the digraph was constructed.
- A (fast) algorithm for the hierarchical classification of districts on the map based on the presence of common borders (neighborhood) between districts was constructed.

Therefore, their further development requires assessments of the stability of the results obtained with variations of these critical levels. In addition, an important role in the development of this topic should be played by estimates of the impact of observation errors on the results obtained in the work. If the array of observations of a system consists of parts of its elements' observations, then in the near future, it will be necessary to develop a procedure for comparing the results of processing these parts in order to determine the most sensitive part.

From the author's point of view, this paper is more applied than theoretical. However, to work with the proposed material, it is required to use a more diverse mathematical tool kit than the one that is traditionally used in the listed applications. In particular, when working with mining materials, this allows us to identify economic and safety indicators and significantly reduce the volume of the analyzed information.

The algorithms presented in this paper appeared as a result of long and rather unsuccessful computational experiments. Practice has shown that in order to obtain reasonable applied results, it is necessary to strictly follow the initial meaningful statement of the problem, but the algorithms proposed by the mathematicians themselves should be convenient in calculations and fast enough. Unfortunately, the consumers of these algorithms are often impatient users.

Funding: This research received no external funding.

Institutional Review Board Statement: Not applicable.

Informed Consent Statement: Not applicable.

Data Availability Statement: This paper has no processing of concrete data.

Conflicts of Interest: The authors declare no conflict of interest.

References

1. Teugels, J.L. *The Class of Subexponential Distributions*; University of Louvain: Annals of Probability: Louvan, Belgium, 1975; Retrieved 7 April 2019.
2. Zolotarev, V.M. *One-Dimensional Stable Distributions*; American Mathematical Society: Providence, RI, USA, 1986.
3. Embrechts, P.; Klueppelberg, C.; Mikosch, T. *Modelling Extremal Events for Insurance and Finance*; Stochastic Modelling and Applied Probability; Springer: Berlin, Germany, 1997; Volume 33.
4. Asmussen, S.R. Steady-State Properties of GI/G/1. *Appl. Probab. Queues* **2003**, *51*, 266–301.
5. Foss, S.; Korshunov, D.; Zachary S. *An Introduction to Heavy-Tailed and Subexponential Distributions*; Springer Science and Business Media: Berlin/Heidelberg, Germany, 2013.
6. Novak, S.Y. *Extreme Value Methods with Applications to Finance*; CRC: London, UK, 2011.
7. Wierman, A. *Catastrophes, Conspiracies, and Subexponential Distributions (Part III). Rigor + Relevance Blog*; RSRG Caltech: Pasadena, CA, USA, 2014.
8. Siebert, C.F.; Siebert, D.C. *Data Analysis with Small Samples and Non-Normal Data Nonparametrics and Other Strategies*; Oxford University Press: Oxford, UK, 2017.
9. Chandrasekharan, S.; Sreedharan, J.; Gopakumar, A. Statistical Issues in Small and Large Sample: Need of Optimum Upper Bound for the Sample Size. *Int. J. Comput. Theor. Stat.* **2019**, *6*. [CrossRef]
10. Konietschke, F.; Schwab, K.; Pauly, M. Small sample sizes: A big data problem in high-dimensional data analysis. *Stat. Methods Med. Res.* **2020**, *30*, 687–701. [CrossRef] [PubMed]
11. Vasileiou, K.; Barnett, J.; Thorpe, S.; Young, T. Characterising and justifying sample size sufficiency in interview-based studies: systematic analysis of qualitative health research over a 15-year period. *BMC Med. Res. Methodol.* **2018**, *18*, 148. [CrossRef] [PubMed]
12. Morgan, C.J. Use of proper statistical techniques for research studies with small samples. *Am. J. Physiol. Lung Cell. Mol. Physiol.* **2017**, *313*, 873–877. [CrossRef] [PubMed]
13. Guzev, M.A.; Rasskazov, I.Y.; Tsitsiashvili, G.S. Algorithm of potentially burst-hazard zones dynamics Representation in massif of rocks by results of seismic-acoustic monitoring. *Procedia Eng.* **2017**, *191*, 36–42. [CrossRef]
14. Zhu, Q.X.; Chen, Z.S.; Zhang, X.H.; Rajabifard, A.; Xu, Y.; Chen, Y.Q. Dealing with small sample size problems in process industry using virtual sample generation: A Kriging-based approach. *Soft Comput.* **2020**, *24*, 6889–6902. [CrossRef]
15. Bulgakov, V.P.; Tsitsiashvili, G.S. Bioinformatics analysis of protein interaction networks: Statistics, topologies, and meeting the standards of experimental biologists. *Biochemistry* **2013**, *78*, 1098–1103. [CrossRef] [PubMed]

16. Tsitsiashvili, G.S.; Bulgakov, V.P.; Losev, A.S. Factorization of Directed Graph Describing Protein Network. *Appl. Math. Sci.* **2017**, *11*, 1925–1931. [CrossRef]
17. Bolotin, E.I.; Tsitsiashvili, G.S.; Golycheva, I.V. Some aspects and perspectives of factor prognosis for the epidemic manifestation of the Tick-Borne Encephalitis based on the multidimensional analysis of temporal rows. *Parazitology* **2002**, *36*, 89–95. (In Russian)
18. Shatilina, T.A.; Tsitsiashvili, G.S.; Radchenkova, T.V. Peculiarities of surface air temperature variations over the Far East regions in 1976–2005. *Russ. Meteorol. Hydrol.* **2010**, *35*, 740–743. [CrossRef]
19. Shatilina, T.A.; Tsitsiashvili, G.S.; Radchenkova, T.V. Okhotsk medium-tropospheric cyclone and its role in the formation of extreme air temperature in January in 1950–2019. *Hydrometeorol. Stud. Forecast.* **2021**, *3*, 64–79. (In Russian) [CrossRef]
20. Tsitsiashvili, G.S.; Shatilina, T.A.; Radchenkova, T.V. *Application of New Algorithms for Processing Meteorological Observations*; Publishing House "Buk": Kazan, Russia, 2022. (In Russian)
21. Lever, J.; Leemput, I.; Weinans, E.; Quax, R.; Dakos, V.; Nes, E.; Bascompte, J.; Scheffer, M. Foreseeing the future of mutualistic communities beyond collapse. *Ecol. Lett.* **2020**, *23*, 2–15. [CrossRef] [PubMed]
22. Radchenko, V. Abundance Dynamics of Pink Salmon, Oncorhynchus gorbuscha, as a Structured Process Determined by Many Factors. *NPAFC Tech. Rep.* **2011**, *8*, 14–18. [CrossRef]
23. Shuntov, V.P.; Temnikh, O.S. *Far Eastern Salmon Industry–2016: Good Results, Successes and Errors in Forecasts and the Traditional Failure of VNIRO on the Ways of Innovative Breakthroughs Announced by Him in Forecasting the Number and Catches of Fish. Study of Pacific salmon in the Far East*; TINRO-Center: Vladivostok, Russia, 2016; Volume 11, pp. 3–13. (In Russian)
24. Rasskazov, I.Y. *Control and Management of Rock Pressure in the Mines of the Far Eastern Region*; Gornaya Kniga Publ.: Moscow, Russia, 2008. (In Russian)
25. Rasskazov, I.Y.; Gladyr, A.V.; Anikin, P.A.; Svyatetsky, V.S.; Prosekin, V.A. Development and modernization of the control system of dynamic appearances of rock pressure on the mines of "Priargunsky Industrial Mining and Chemical Union". *JSC Gorn. Zhurnal (Min. J.)* **2013**, *8*, 9–14. (In Russian)
26. Tsitsiashvili, G.S.; Bulgakov, V.P.; Losev, A.S. Hierarchical classification of directed graph with cyclically equivalent nodes. *Appl. Math. Sci.* **2016**, *10*, 2529–2536.
27. Mezic, I.; Fonoberov, V.A.; Fonoberova, M.; Sahai, T. Spectral Complexity of Directed Graphs and Application to Structural Decomposition. *Complexity* **2019**, *2019*, 9610826. [CrossRef]
28. Tarjan, R. Depth-first Search and Linear Graph Algorithms. *SIAM J. Comput.* **1972**, *1*, 146–160. [CrossRef]
29. Pikunov, D.G.; Mikell, D.G.; Seredkin, I.V.; Nikolaev, I.G.; Dunishenko, Y.M. *Winter Tracking Records of the Amur Tiger in the Russian Far East (Methodology and History of Accounting)*; Dalnauka: Vladivostok, Russia, 2014. (In Russian)
30. Kingman, J.F.C. *Poisson Processes*; Oxford Studies in Probability-3; Clarendon Press: Oxford, UK, 1993.

 *mathematics*

MDPI

Article

The *Geo*/$G^{a,Y}$/1/N Queue Revisited

Mohan Chaudhry [1] and Veena Goswami [2],*

[1] Department of Mathematics and Computer Science, Royal Military College of Canada, P.O. Box 17000, Kingston, ON K7K 7B4, Canada
[2] School of Computer Applications, Kalinga Institute of Industrial Technology, Bhubaneswar 751 024, India
* Correspondence: veena@kiit.ac.in

Abstract: We not only present an alternative and simpler approach to find steady-state distributions of the number of jobs for the finite-space queueing model $Geo/G^{a,Y}/1/N$ using roots of the inherent characteristic equation, but also correct errors in some published papers. The server has a random service capacity Y, and it processes the jobs only when the number of jobs in the system is at least 'a', a threshold value. The main advantage of this alternative process is that it gives a unified approach in dealing with both finite- and infinite-buffer systems. The queue-length distribution is obtained both at departure and random epochs. We derive the relation between the discrete-time Geo/$G^{a,Y}$/1/N queue and its continuous-time analogue. Finally, we deal with performance measures and numerical results.

Keywords: batch service; roots; discrete-time queue; discrete renewal theory; finite buffer capacity

MSC: 60K25; 68M20; 90B22

1. Introduction

Discrete-time queues with batch service have numerous applications in various areas such as transportation systems, traffic, production, manufacturing, telecommunication, and cloud computing. In various real-life settings, it is often noted that the jobs are served in batches. The server may serve with fixed maximum or variable capacity in batch service systems. For more details on batch service queues, one may refer to Chaudhry and Templeton [1] as well as Medhi [2]. Discrete-time queues are more notable in systems' modelling, see [3–5].

In discrete-time queues, it is assumed that arrivals and departures occur at boundary epochs of time slots. Further, discrete-time queues deal with an early arrival system (EAS) or a late arrival system with delayed access (LAS-DA). For more on this, see Hunter [4]. We may note that EAS and LAS-DA policies are similar to departure-first (DF) and arrival-first (AF), respectively; see Gravey and Hébuterne [6].

Several researchers study single-server batch-service discrete-time queues with various phenomena, see [7–13]. In Gupta and Goswami [10], they discuss a discrete-time finite-buffer general bulk service queue under both LAS-DA and EAS policies. The model involving batch-size-dependent service in a discrete-time queue where inter-arrival times and the service times follow geometric and general distribution, respectively, has been discussed by Banerjee et al. [14]. In Yi et al. [13], the authors analyze a discrete-time $Geo/G^{a,Y}/1/N$ queue, where service is initiated only when the number of jobs in the system is at least 'a'. In Zeng and Xia [15], the authors discuss M/$G^{a,b}$/1/N queue where service is in batches with minimum threshold a, maximum capacity b and the buffer size, N, finite or infinite.

At some point, finding the roots of the characteristic equation seemed difficult, mainly when the number involved was large. Several researchers have made these comments. Given this, the procedure for solving queueing models led to the matrix-analytic or matrix-geometric method. In this connection, see the remark below. Following Chaudhry et al. [16],

Citation: Chaudhry, M.; Goswami, V. The *Geo*/$G^{a,Y}$/1/N Queue Revisited. *Mathematics* **2022**, *10*, 3142. https://doi.org/10.3390/math10173142

Academic Editors: Gurami Tsitsiashvili and Alexander Bochkov

Received: 25 July 2022
Accepted: 23 August 2022
Published: 1 September 2022

Publisher's Note: MDPI stays neutral with regard to jurisdictional claims in published maps and institutional affiliations.

other researchers have attempted to show that roots can be found, and using them leads to nice analytic and numerical solutions. Gouweleeuw [17] shows that the roots' approach for finding probabilities from generating functions in precise expressions is effective. Further, in the case of large buffer size, solving simultaneous equations gives rise to poor reliability and takes considerable time; see Powell [18] (p. 141), who, while dealing with the model $M/D^c/1$, states that when $c \geq 40$ the method using simultaneous equations breaks down leading to negative probabilities. The goal of this paper is to give an alternative solution that is analytically powerful, simple, and easy to implement numerically. It may be stated further that this method has not been used to discuss the discrete-time queueing system that deals with batch services and a finite buffer.

In real-world systems, we encounter many finite-buffer systems such as telecommunication networks. Because of this, we study the $Geo/G^{a,Y}/1/N$ model under the assumption of late arrival and delayed access system (LAS-DA). Here, we assume that the single server with variable service capacity will process the jobs only when there are at least 'a' jobs in the system. In Yi et al. [13], the authors found the queue-length distribution at post-departure by solving simultaneous equations and random epoch by applying the "rate in = rate out" arguments. We develop an alternate process to find the queue-length distributions at post-departure and random epochs.

The principal contributions of this work may be summed up as follows:

- We find an alternative method to obtain the steady-state queue-length distributions of $Geo/G^{a,Y}/1/N$ at post-departure and random epochs.
- The approach presented in this paper unifies in a way that can handle both the infinite-space as well as finite-space models at the same time.
- We point out the incorrectness of queue-length distributions' numerical results (at random epochs) reported in Yi et al. [13]. They also assumed batches with a random capacity Y having probability mass function (pmf) $P(Y = i) = y_i, i = 0, 1, \ldots, b$ instead of $i = a, a+1, \ldots, b$.
- We compute the steady-state queue-length distributions of $Geo/G^{a,Y}/1/N$ at post-departure and random epochs when $\rho > 1$, which is missing in Yi et al. [13]. Further, we point out the incorrectness of the formula for the mean waiting time in the queue (using Little's rule) in Yi et al. [13].
- We can obtain the continuous-time solution for the model $M/G^{a,Y}/1/N$ (see Appendix A) and the procedure used here can be applied to obtain a solution for this continuous-time model too. Further, it is anticipated that, using this method, we can obtain waiting-time distribution using Little's law, a problem for which no solution is available, even using the matrix-analytic method. The primary purpose of this paper is to show its unifying power and superiority over other methods, and to give a simple solution to the existing problem.
- Finally, we compare the roots' method against the process that uses simultaneous equations and present the results in the numerical section. It clearly shows that the roots approach takes less time.

The remaining paper is structured as follows. Section 2 specifies the model. Section 3 analyzes the $Geo/G^{a,Y}/1/N$ system and finds queue-length distributions for the LAS-DA policy. Section 4 examines various system performance measurements. Section 5 provides some numerical results and, finally, the paper is concluded in Section 6.

Remark 1. *It may be useful to comment on the matrix-analytic and the root-finding method. Kendall [19] made a famous remark that queueing theory wears the Laplacian curtain. Kleinrock [20] (p. 291) states, "One of the most difficult parts of this method of spectrum factorization is to solve for the roots". Neuts (see Neuts' book [21] and also Stidham [22]) states, "In discussing matrix-analytic solutions, I had pointed out that when the Rouch' roots coincide or are close together, the method of roots could be numerically inaccurate. When I finally got copies of Crommelin's papers, I was elated to read that, for the same reasons as I, he was concerned about the clustering of roots. In 1932, Crommelin knew; in 1980, many of my colleagues did not . . . ". Following this, several other* *researchers made similar comments. Given this, the procedure for solving queueing models led to the matrix-analytic or matrix-geometric method. In response, Chaudhry et al. [16] showed that the roots can be found even when the number involved is large. (This was done when MATHEMATICA OR MAPLE failed to give those roots. We can now use this software to find roots.)*

2. Model Description

We consider a $Geo/G^{a,Y}/1/N$ queue where jobs arrive following a Bernoulli process with parameter λ. A single server processes the jobs on a first-come-first-served (FCFS) discipline in batches with a random capacity Y possessing a probability mass function (pmf) $P(Y = i) = y_i, i = a, a+1, \ldots, b$ with $\sum_{i=a}^{b} y_i = 1$, the probability generating function (PGF) $Y(z) = \sum_{i=a}^{b} y_i z^i$, mean $E(Y) = \sum_{i=a}^{b} i y_i$ and $Y'(z) = z^b Y(z^{-1})$. We assume the minimum and maximum threshold values of the random variable Y as a and b, respectively. When there are at least 'a' jobs in the queue, the server commences serving a batch of size i with probability y_i (when there are $Y \leq b$, it takes all of them). When the number of jobs comes down below a threshold value $a(\geq 1)$ in the system, the server remains idle but awaits the number of jobs to rise to a; when it attains a, it resumes service. The service times $\{S_n, n \geq 1\}$ are independently and identically distributed (iid) with arbitrary pmf $s_k = P(S_n = k), k = 1, 2, \ldots$ and $s_0 = 0$, the PGF $S(z) = \sum_{i=1}^{\infty} s_i z^i$ and mean service time $E(S) = s = \frac{d}{dz} S(z)|_{z=1} = 1/\mu$.

The processing times of the jobs are independent of the arrival process and the number of jobs served. The waiting buffer has a finite capacity N with $b \leq N$. Thus, in the system, no more than $N + b$ jobs can be available anytime. We presume offered load of the system as $\rho = \lambda E(S)/E(Y)$. In LAS-DA policy, arrivals occur in $(u-, u)$, and departures take place in $(u, u+)$; arrivals supersede departures. Figure 1 describes the various time periods at which events occur. For more details on this, one may refer to [5,6].

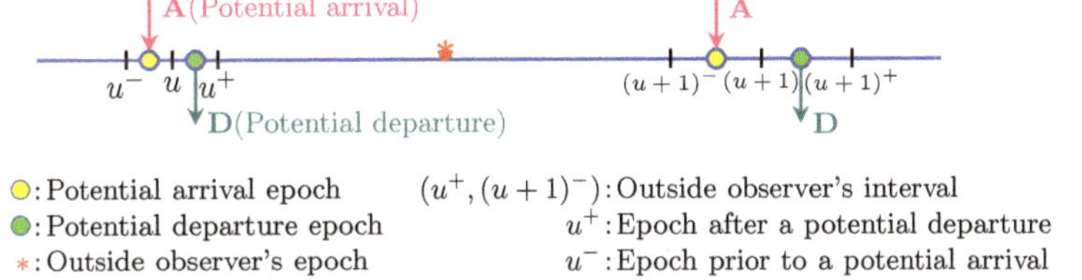

○ : Potential arrival epoch $(u^+, (u+1)^-)$: Outside observer's interval
● : Potential departure epoch u^+ : Epoch after a potential departure
✳ : Outside observer's epoch u^- : Epoch prior to a potential arrival

Figure 1. Various time epochs in LAS-DA.

3. Queue-Length Distributions

Here, we find steady-state queue-length distributions at various epochs of $Geo/G^{a,Y}/1/N$ queue with the LAS-DA policy.

3.1. Post-Departure Epoch Probabilities

Let Q_ℓ^+ be the number of jobs in the queue after completing the ℓth service. Suppose $A_{\ell+1}$ and $Y_{\ell+1}$ represent the number of arrivals throughout the processing time on the $(\ell+1)$th job and the processing capacity of $(\ell+1)$th service, respectively. As per the batch service rule, the departure epoch queue lengthis

$$Q_{\ell+1}^+ = \min\left((Q_\ell^+ - Y_{\ell+1})^+ + A_{\ell+1}, N \right),$$

where $x^+ = \max(x, 0)$. The probability distribution of $A_{\ell+1}$ is

$$k_n = P(A_{\ell+1} = n) = \sum_{j=1}^{\infty} P(A_{\ell+1} = n | S_{\ell+1} = j) P(S_{\ell+1} = j)$$

$$= \sum_{j=n}^{\infty} s_j \binom{j}{n} \lambda^n (1-\lambda)^{j-n}, \ n \geq 0.$$

Here we may note that arrivals are generated by a Bernoulli sequence by the property of geometric interarrival times. In LAS-DA, if the service time of the $(\ell+1)$th job is j slots, then there will be j time slots where n arrivals may occur. One may note that $\binom{j}{n}\lambda^n(1-\lambda)^{j-n}$ is the probability of n arrivals in j slots. Let $K(z) = \sum_{n=0}^{\infty} k_n z^n$ be the probability generating function of the sequence $\{k_n, \ n = 0, 1, \ldots\}$. Thus

$$K(z) = \sum_{n=0}^{\infty} \sum_{j=n}^{\infty} s_j \binom{j}{n} \lambda^n (1-\lambda)^{j-n} z^n = \sum_{j=0}^{\infty} s_j (1 - \lambda + \lambda z)^j = S(1 - \lambda + \lambda z).$$

Transition probabilities in one step of underlying Markov chain for $p_{ij} = \lim_{\ell \to \infty} Pr\{Q^+_{\ell+1} = j | Q^+_\ell = i\}$ are given as

$$p_{ij} = \begin{cases} k_j, & 0 \leq i \leq a, \ 0 \leq j \leq N-1, \\[2mm] k_j \sum_{r=i}^{b} y_r + \sum_{r=a}^{i-1} y_r k_{j-i+r}, & a+1 \leq i \leq b-1, \ i-a-1 \leq j \leq N-1, \\[2mm] \sum_{r=a}^{b} y_r k_{j-i+r}, & b \leq i \leq N, \ i-b \leq j \leq N-1, \\[2mm] \ell_N, & 0 \leq i \leq a, \ j = N, \\[2mm] \ell_N \sum_{r=i}^{b} y_r + \sum_{r=a}^{i-1} y_r \ell_{N-i+r}, & a+1 \leq i \leq b-1, \ j = N, \\[2mm] \sum_{r=a}^{b} y_r \ell_{N-i+r}, & b \leq i \leq N, \ j = N, \end{cases} \tag{1}$$

where $\ell_j = \sum_{r=j}^{\infty} k_r$ and k_j with $j < 0$ defined to be zero, and which leads to the transition probability matrix $P = (p_{ij})$ as

$$P = \begin{array}{c} \\ 0 \\ 1 \\ \vdots \\ a \\ a+1 \\ \\ \vdots \\ b \\ b+1 \\ \vdots \\ N \end{array}
\begin{array}{c}
\begin{array}{ccccccc} 0 & 1 & \cdots & j & \cdots & N-1 & N \end{array} \\
\left(\begin{array}{ccccccc}
k_0 & k_1 & \cdots & k_j & \cdots & k_{N-1} & \ell_N \\
k_0 & k_1 & \cdots & k_j & \cdots & k_{N-1} & \ell_N \\
\vdots & \vdots & \vdots & \vdots & \ddots & \vdots & \vdots \\
k_0 & k_1 & \vdots & k_j & \cdots & k_{N-1} & \ell_N \\
k_0 \sum_{r=a+1}^{b} y_r & k_1 \sum_{r=a+1}^{b} y_r & \vdots & k_j \sum_{r=a+1}^{b} y_r & \cdots & k_{N-1}\sum_{r=a+1}^{b} y_r & \ell_N \sum_{r=a+1}^{b} y_r \\
+k_0 y_a & +k_{j-1} y_a & & & & +k_{N-2} y_a & +\ell_{N-1} y_a \\
\vdots & \vdots & \vdots & \ddots & \vdots & \vdots & \vdots \\
k_0 y_b & \sum_{r=b-1}^{b} y_r k_{1-b+r} & \vdots & \sum_{r=a}^{b} y_r k_{j-b+r} & \cdots & \sum_{r=a}^{b} y_r k_{N-1-b+r} & \sum_{r=a}^{b} y_r k_{N-b+r} \\
0 & k_0 y_b & \vdots & \sum_{r=a}^{b} y_r k_{j-b+r-1} & \cdots & \sum_{r=a}^{b} y_r k_{N-b+r-1} & \sum_{r=a}^{b} y_r k_{N-b+r-1} \\
\vdots & \vdots & \vdots & \ddots & \vdots & \vdots & \vdots \\
0 & 0 & 0 & \sum_{r=a}^{b} y_r k_{j-N+r} & \vdots & \sum_{r=a}^{b} y_r k_{r-1} & \sum_{r=a}^{b} y_r \ell_r
\end{array}\right)
\end{array}$$

The $(\ell+1)$th batch service starts when there are 'a' jobs in the queue, and the state changeover occurs from $0 \le i \le a-1$ to $0 \le j \le N-1$. If the state changeover is from state $i \ge a$ to state $0 \le j \le N-1$, then there is a busy period between the leaving epoch of the ℓth batch and the commencement of processing $(\ell+1)$st batch. Let the steady-state probability $p^+ = \{P_0^+, P_1^+, \ldots, P_n^+, \ldots, P_N^+\}$ represent ℓ jobs at departure epochs. Then, in steady-state, $p^+ = p^+ P$ can be expressed as follows:

$$P_j^+ = \sum_{i=0}^{a} P_i^+ k_j + \sum_{i=a+1}^{b-1} P_i^+ \sum_{r=i}^{b} y_r k_j + \sum_{i=a+1}^{b-1} P_i^+ \sum_{r=a}^{i-1} y_r\, k_{j-i+r} + \sum_{i=b}^{N} P_i^+ \sum_{r=a}^{b} y_r\, k_{j-i+r},$$

$$0 \le j \le N-1, \tag{2}$$

$$P_N^+ = \sum_{i=0}^{a} P_i^+ \ell_N + \sum_{i=a+1}^{b-1} P_i^+ \sum_{r=i}^{b} y_r\ell_N + \sum_{i=a+1}^{b-1} P_i^+ \sum_{r=a}^{i-1} y_r\, \ell_{N-i+r} + \sum_{i=b}^{N} P_i^+ \sum_{r=a}^{b} y_r\, \ell_{N-i+r}, \tag{3}$$

where the normalization condition is $\sum_{j=0}^{N} P_j^+ = 1$. It may be noted that Equation (3) is redundant and will not be considered in analysis hereafter. Specify PGF of P_j^+ as $P^+(z) = \sum_{j=0}^{N} P_j^+ z^j$. Multiplying Equation (2) by z^j and then adding overall j, we obtain

$$P^+(z) = \sum_{j=0}^{N-1} k_j z^j \sum_{i=0}^{a} P_i^+ + \sum_{j=0}^{N-1} z^j \sum_{i=a+1}^{b-1} P_i^+ \left(\sum_{r=i}^{b} y_r\right) k_j$$

$$+ \sum_{j=0}^{N-1} z^j \sum_{i=a+1}^{b-1} P_i^+ \sum_{r=a}^{i-1} y_r\, k_{j-i+r} + \sum_{j=0}^{N-1} z^j \sum_{i=b}^{N} P_i^+ \sum_{r=a}^{b} y_r\, k_{j-i+r} + P_N^+ z^N,$$

$$P^+(z)\left[1 - K(z)Y\left(\frac{1}{z}\right)\right] = K(z) \sum_{i=0}^{a-1} P_i^+ \left(1 - z^i Y\left(\frac{1}{z}\right)\right) + P_N^+ z^N$$

$$+ K(z) \sum_{i=a}^{b} P_i^+ \left(\sum_{r=i}^{b} y_r - z^i \sum_{r=i}^{b} \frac{y_r}{z^r}\right) - \sum_{i=a}^{b} y_i \sum_{j=i+1}^{N} P_j^+ \sum_{r=N-j+a}^{\infty} k_r z^{r+j-i}$$

$$- \left(\sum_{i=0}^{a-1} P_i^+ + \sum_{i=a}^{b} P_i^+ \sum_{r=i}^{b} y_r\right) \sum_{j=N}^{\infty} k_j z^j.$$

Simplifying the above equation, we get

$$P^+(z) = \frac{K(z)\left[\sum_{i=0}^{a-1} P_i^+ \left(z^b - z^i Y'(z)\right) + \sum_{i=a}^{b-1} P_i^+ \left(\sum_{r=i}^{b} y_r z^b - z^i \sum_{r=i}^{b} y_r z^{b-r}\right)\right]}{z^b - K(z)Y'(z)}$$

$$+ \frac{z^{N+b}\left(P_N^+ - \sum_{i=a}^{b} y_i \sum_{j=i+1}^{N} P_j^+ \sum_{r=N-j+a}^{\infty} k_r z^{r+j-i-N}\right)}{z^b - K(z)Y'(z)}$$

$$- \frac{z^{N+b}\left(\sum_{i=0}^{a-1} P_i^+ + \sum_{i=a}^{b} P_i^+ \sum_{r=i}^{b} y_r\right) \sum_{j=N}^{\infty} k_j z^{j-N}}{z^b - K(z)Y'(z)}. \tag{4}$$

Only the first expression on the right side of the Equation (4) will add to the coefficient of z^j, $j = 0, 1, \ldots N$. To the right of Equation (4), we ignore the second and third expressions, which consist of an output of z higher than $N + b$. These are not required as we want to compare the coefficients of z^j for $j \le N$ on both sides in Equation (4) to find P_j^+ for $j = 0, 1, \ldots, N$. Let

$$P_N^+(z) = \frac{K(z)\left[\sum\limits_{i=0}^{a-1} P_i^+\left(z^b - z^i Y'(z)\right) + \sum\limits_{i=a}^{b-1} P_i^+\left(\sum\limits_{r=i}^{b} y_r z^b - z^i \sum\limits_{r=i}^{b} y_r z^{b-r}\right)\right]}{z^b - K(z)Y'(z)}, \tag{5}$$

which is equivalent to the PGF of an infinite buffer case. The function $P_N^+(z)$ is fully determined once P_i^+, $i = 0, 1, \ldots, b-1$ are known. One may observe that when $\rho < 1$, Equation (5) denotes the PGF of discrete-time infinite buffer $Geo/G^{a,Y}/1$ queue. We can calculate the probabilities for $\rho \ge 1$ in the case of a finite buffer $Geo/G^{a,Y}/1/N$ queue.

Remark 2. *Using $a = 1$, the model is reduced to $Geo/G^Y/1/N$ queue.*

Remark 3. *Taking $y_1 = 1$, $y_i = 0$, $\forall\, 2 \le i \le b$ and $a = 1$, the model becomes $Geo/G/1/N$ queue and Equation (5) establishes PGF as $P_N^+(z) = \frac{P_0^+ K(z)(1-z)}{K(z)-z}$, where $K(z) = S(1 - \lambda + \lambda z)$, which corresponds to the results of [23].*

Remark 4. *Taking $y_b = 1$ and $y_i = 0$, $\forall\, i \neq b$, the model is reduced to $Geo/G^{(a,b)}/1/N$ and Equation (5) establishes the PGF as $P_N^+(z) = \frac{K(z)\sum\limits_{i=0}^{b-1} P_i^+\left(z^b - z^i\right)}{z^b - K(z)}$.*

Intending to establish a unified method to solve the queueing system $Geo/G^{a,Y}/1/N$, we obtain $\{P_n^+\}_0^N$ from $P^+(z)$ by using the roots of characteristic equation and partial-fraction expansion. The literature on queueing systems shows that arrival/service-time distributions that possess the generating function as a rational function deal with the broad class of distributions see [24]. For this, we suppose that $K(z) = S(1 - \lambda + \lambda z)$ as a rational function in z, specified by

$$K(z) = S(1 - \lambda + \lambda z) = \frac{f(z)}{g(z)},$$

where $f(z)$ and $g(z)$ are polynomials of degree m and s, respectively, where m and s can have any value, e.g., m can be greater than s, e.g., see Example 3(ii). Thus, we have from Equation (5),

$$P_N^+(z) = \frac{f(z)\left[\sum\limits_{i=0}^{a-1} P_i^+\left(z^b - z^i Y'(z)\right) + \sum\limits_{i=a}^{b-1} P_i^+\left(\sum\limits_{r=i}^{b} y_r z^b - z^i \sum\limits_{r=i}^{b} y_r z^{b-r}\right)\right]}{z^b g(z) - f(z)Y'(z)}. \tag{6}$$

The denominator of Equation (6) is a polynomial of degree $b + s$ which when equated to zero has $b + s$ roots inside, on, or outside the unit circle $\mid z \mid = 1$, say, $\gamma_1 = 1$, γ_i $(i = 2, 3, \ldots, b+s)$.

Remark 5. *If the denominator $z^b g(z) - f(z)Y'(z)$ of Equation (6) $= 0$ obtains roots close to each other or repeated roots, we may obtain them by applying advanced software packages, for instance, MATHEMATICA or MAPLE. The MAPLE script for calculating repeated roots is exemplified below for Equation*

$$u(y) = (y-2)(y-5)^2(y-7)^3(y-11).$$

$restart : Digits := 10 : with(RootFinding) :$

$m := (y-2)(y-5)^2(y-7)^3(y-11);$

$Analytic(u, y, re = -1..10, im = -2..10);$

7.00000000000000, 7.00000000000000, 7.00000000000000, 5.00000000000000,

5.00000000000000, 2.00000000000000, 11.00000000000000

According to Rouché's theorem, the denominator of Equation (6) has b zeros say, γ_i $(i = 1, 2, \ldots, b)$ inside the unit circle. As $P_N^+(z)$ converges in $\mid z \mid \leq 1$, the b zeros within the unit circle of the denominator should cancel with the b zeros of the numerator. After canceling the b factors in the numerator and denominator, we can re-write Equation (6) as

$$P_N^+(z) = T\left(C(z) + \frac{f_1(z)}{\prod\limits_{i=b+1}^{b+s}(z-\gamma_i)}\right), \tag{7}$$

where $C(z) = \sum\limits_{i=0}^{n_0} c_i z^i$ and T is a normalizing constant. Note that when $2m < s$, then $C(z)$ will be zero. In the partial-fraction process, a slight modification is needed ([25], p. 221) when all the roots are not distinct. Since we are looking at the finite buffer queue system, three instances appear here.

- If $\rho < 1$ the s roots γ_i, $i = b+1, b+2, \ldots, b+s$ remain outside the circle $\mid z \mid = 1$.
- If $\rho = 1$, among the s roots, one root is '1', and the other roots γ_i, $i = b+2, b+3, \ldots, b+s$ are outside the unit circle $\mid z \mid = 1$.
- If $\rho > 1$, among s roots, one root is inside, say γ_{b+1} and the other roots γ_i, $i = b+2, b+3, \ldots, b+s$ are outside, see [26]. One may note that when $\rho > 0$ increases, one positive real root comes closer to the origin from right to left.

The expression (7) is tractable for inversion. Applying partial-fraction expansion to Equation (7) yields

$$P_N^+(z) = T\left(C(z) + \sum_{i=b+1}^{b+s}\frac{M_i}{z-\gamma_i}\right), \tag{8}$$

where

$$M_i = \frac{f_1(\gamma_i)}{\prod\limits_{j=b+1, j\neq i}^{b+s}(\gamma_i - \gamma_j)}.$$

Using Equation (8), we have

$$P_n^+ = \begin{cases} T\left(c_n + \sum\limits_{i=b+1}^{b+s}\frac{-M_i}{\gamma_i^{n+1}}\right), & \text{if } n = 0, 1, \ldots, n_0, \\[2em] T\sum\limits_{i=b+1}^{b+s}\frac{-M_i}{\gamma_i^{n+1}}, & \text{if } n = n_0+1, n_0+2, \ldots, N. \end{cases} \tag{9}$$

Employing the normalization condition $\sum_{n=0}^{N} P_n^+ = 1$, we obtain the only unknown T as

$$
T = \begin{cases} \left(\sum_{i=0}^{n_0} c_i - \sum_{i=b+1}^{b+s} \frac{M_i}{\gamma_i} \times \frac{1-\gamma_i^{-(N+1)}}{1-\gamma_i^{-1}} \right)^{-1}, & \text{if } \rho \neq 1, \\[4ex] \left(\sum_{i=0}^{n_0} c_i - M_{b+1}(N+1) - \sum_{i=b+2}^{b+s} \frac{M_i}{\psi_i} \times \frac{1-\psi_i^{-(N+1)}}{1-\psi_i^{-1}} \right)^{-1}, & \text{if } \rho = 1. \end{cases}
\tag{10}
$$

Thus, once all the roots are known, we can get the distribution for the number in queue.

Remark 6. *It may be noted that it is also possible to find the probabilities $\{P_n^+\}_0^N$ by assuming the solution of the form $P_j^+ = Cz^j$, where $C \neq 0$. Unfortunately, if we use this method, we have to solve for N simultaneous equations.*

3.2. Relationship between the Queue-Length Distributions at Post-Departure and Random Epochs

This sub-section establishes associations between probability at random and post-departure epochs by basic probabilistic reasoning and discrete-time renewal theory. In steady-state, let $\{P_j\}_0^N$ and $\{P_j^-\}_0^N$ be the probabilities representing the number of jobs in the queue at random times and before arrival, respectively. Since the inter-arrival times use geometric distribution, the arrivals are independent of other events. Thus, it implies that $P_j = P_j^-, \forall j = 0, 1, \ldots, N$; for details, see Boxma and Groenendijk [27]. If the server is idle, there are $< a$ jobs in the queue. Suppose N_q is the number of tasks in the queue at some random time. At a random epoch, the steady-state probabilities are $P_{n,0} = P(N_q = n,$ server idle), $0 \leq n \leq a - 1$, and $P_{n,1} = P(N_q = n,$ server busy), $0 \leq n \leq N$. Given this,

$$
P_j = \begin{cases} P_{j,0} + P_{j,1} & \text{if } 0 \leq j \leq a - 1 \\ P_{j,1} & \text{if } a \leq j \leq N \end{cases}
$$

Let the limiting pmf of the elapsed service time be $\hat{s}_\ell$, which is determined by $\hat{s}_\ell = \mu \sum_{r=\ell+1}^{\infty} s_r, \ell \geq 0$ (see, [5] (p. 20)), and $\hat{k}_\ell$ be the probability that the number of arrivals during an elapsed service time is ℓ. This yields

$$
\hat{k}_\ell = \sum_{i=\ell}^{\infty} \binom{i}{\ell} \lambda^\ell (1-\lambda)^{i-\ell} \hat{s}_i, \quad \ell = 0, 1, \ldots
$$

If E^* is the mean inter-departure time of processing batches, $1/E^*$ represents the departure rate. At the batch departure epoch, if the number of jobs in the queue is less than a, the subsequent batch departure occurs after an idle time and the service time processing time. Otherwise, the release of the next batch takes place following the processing time of the subsequent batch. This gives,

$$
E^* = E(S)\left(1 - \sum_{i=0}^{a-1} P_i^+\right) + \sum_{i=0}^{a-1} P_i^+\left(\frac{a-i}{\lambda} + E(S)\right) = E(S) + \sum_{i=0}^{a-1} P_i^+ \frac{(a-i)}{\lambda}.
$$

Remark 7. *It can also be put down as $E^* = E(S) + \sum_{i=1}^{a} P_{a-i}^+\left(\frac{i}{\lambda}\right)$. When $a = 1$, it matches with Chaudhry and Chang [7].*

Theorem 1. *The random- and post-departure-epoch probabilities* $\{\{P_{j,0}\}_0^{a-1}, \{P_{j,1}\}_0^N\}$, *and* $\{P_j^+\}_0^N$ *are related by*

$$P_{j,0} = \frac{\sum_{i=0}^{j} P_i^+}{\sum_{i=0}^{a-1} (a-i)P_i^+ + \lambda E(S)}, \quad 0 \leq j \leq a-1, \tag{11}$$

$$P_{j,1} = \left(1 - \sum_{i=0}^{a-1} P_{i,0}\right) \cdot \left[\sum_{i=0}^{a-1} P_i^+ \hat{k}_j + \sum_{i=a}^{b-1} P_i^+ \left(\sum_{m=i}^{b} y_m\right)\hat{k}_j + \sum_{i=a+1}^{b-1} P_i^+ \sum_{m=a}^{i-1} y_m \, \hat{k}_{j-i+m} \right.$$
$$\left. + \sum_{i=b}^{N} P_i^+ \sum_{m=a}^{b} y_m \, \hat{k}_{j-i+m} \right], \quad 0 \leq j \leq N-1. \tag{12}$$

Finally, $P_{N,1}$ *can be found from* $P_{N,1} = 1 - \sum_{j=0}^{a-1} P_{j,0} - \sum_{j=0}^{N-1} P_{j,1}$.

Proof. The fraction of the time the batch server remains idle between two successive departure epochs is the probability of getting the server idle at a random epoch (P_{idle}). Let $E(I)$ be the mean idle period. Using the definition of $E(S)$ and $E(I)$, we have

$$P_{idle} = \frac{E(I)}{E(I) + E(S)},$$

$$P_{j,0} = P_{idle} \times P(\text{fraction of idle period}) = \frac{E(I)}{E(I) + E(S)} \times \frac{\frac{1}{\lambda}\sum_{i=0}^{j} P_i^+}{E(I)}, \quad 0 \leq j \leq a-1,$$

where $E(I) = \frac{1}{\lambda}\sum_{i=0}^{a-1} P_i^+ (a-i)$. We employ system state conditioning and discreet renewal theory to find $P_{j,1}$. The processor is active with probability $(1 - \sum_{i=0}^{a-1} P_{i,0})$; thus,

$$P_{j,1} = P(N_q = j, \text{processor active}) = (1 - \sum_{i=0}^{a-1} P_{i,0})P(N_q = j \mid \text{processor active}) \tag{13}$$

Assuming that k_j^e is the number of jobs that come following an embedded Markov point until the elapsed service time, we have

$$P(N_q = j \mid \text{processor active}) = \sum_{i=0}^{a-1} P_i^+ \sum_{m=a}^{b} y_m \hat{k}_j + \sum_{i=a}^{b-1} P_i^+ \sum_{m=i}^{b} y_m \hat{k}_j$$
$$+ \sum_{i=a+1}^{b-1} P_i^+ \sum_{m=a}^{i-1} y_m \, \hat{k}_{j-i+m} + \sum_{i=b}^{N} P_i^+ \sum_{m=a}^{b} y_m \, \hat{k}_{j-i+m},$$
$$0 \leq j \leq N-1. \tag{14}$$

Putting together (13) and (14), we obtain (12). We can obtain $P_{N,1}$ using normalization condition. Thus, we obtain random epoch probabilities $\{\{P_{j,0}\}_0^{a-1}, \{P_{j,1}\}_0^N\}$ in connection with post departure epoch probabilities $\{P_j^+\}_0^N$. $\square$

Though the relations between random- and post-departure epoch probabilities are available in [13] using transition rates, here, we develop an alternative method to obtain the queue-length distributions at random epochs.

Because of BASTA (Bernoulli arrivals see time averages) property, see [5], the queue-length distribution exactly before arrival of job will be equal to that of $P_{j,0}$ and $P_{j,1}$. Further, since the outside observation epoch falls in an interval between arrival and departure epochs, the outside observer's distribution is the same as the random epoch distribution.

4. Performance Measures

This section deals with several measures of performance. The average number of jobs in the queue (L_q) is given by

$$L_q = \sum_{j=0}^{a-1} j\, P_{j,0} + \sum_{j=0}^{N} j\, P_{j,1}$$

The probability of the processor being busy (PB) at some random moment is specified by $1 - \sum_{j=0}^{a-1} P_{j,0}$. Due to the BASTA property, the loss or blocking probability (PBL) is given by $PBL = P_{N,1}$. Since the effective arrival rate $\lambda_e = \lambda(1 - P_{N,1})$, we can obtain the average wait time in the queue (W_q) by employing Little's law as $W_q = \frac{L_q}{\lambda_e}$. The reported result of W_q in [13] is incorrect. They have applied the effective arrival rate as λ instead of λ_e.

Remark 8. *If $\rho < 1$ and $N \to \infty$, then $\lambda_e = \lambda$ and $L_q = \sum\limits_{j=0}^{a-1} j\, P_{j,0} + \sum\limits_{j=0}^{\infty} j\, P_{j,1}$. Using Little's law, the average waiting time in the queue (W_q) can be computed as $W_q = \frac{L_q}{\lambda}$.*

5. Numerical Results

To exemplify the analytic results found in this article, we illustrate several numerical outcomes in tables and figures. We also give several performance measures, for instance, the average queue length (L_q), the average waiting time in the queue (W_q), and the probability of loss (P_{loss}). The computations were performed in double precision and executed in a 64-bit windows 10 professional OS possessing Intel(R) corei5-6200U processor @2.30 GHz and 8 GB DDR3 RAM utilizing MAPLE 18 software. The numerical results were exact, but we reported outcomes rounding to six decimal places. Because reported outcomes are rounded, the sum of probabilities may not add up to one in some cases.

Example 1. *$Geo/NB_2^{a,Y}/1/10$queue. Consider the distribution of service time as being in two stage negative binomial distribution (NB) with PGF $S(z) = \left(\frac{\mu z}{1 - \beta z}\right)^2$. Here, we consider the same parameters as in Table 1 of the paper [13] to compare the results. The parameters are $E(S) = 5$, $y_2 = 0.2$, $y_3 = 0.1$, $y_4 = 0.7$, and $E(Y) = 3.5$. The arrival rates are 0.14, 0.35, and 0.56 with corresponding traffic intensities $\rho = 0.2$, 0.5, and 0.8, respectively. To show the evaluation process, let us assume $\lambda = 0.14$. The denominator of Equation (6) has six roots, two of which are outside, and four are in and on the unit circle. From Equation (8),*

$$P_N^+(z) = T\left(0.068061 + \frac{7.817631}{z - 6.392624} - \frac{6.204397}{z - 5.024572}\right),$$

where $T = 6.250001$. Similarly, the denominator of $P_N^+(z)$ has two roots outside the unit circle when arrival rates are 0.35 and 0.56, and we have from Equation (8),

$$P_N^+(z) = T\left(c_0 + \frac{M_{b+1}}{z - \gamma_{b+1}} + \frac{M_{b+2}}{z - \gamma_{b+2}}\right),$$

where $\gamma_{b+1} = 2.106094$, $\gamma_{b+2} = 3.454495$, $T = 6.256038$, $c_0 = 0.053212$, $M_{b+1} = -0.619857$, $M_{b+2} = 1.113392$ and $\gamma_{b+1} = 1.304299$, $\gamma_{b+2} = 2.688401$, $T = 6.867372$, $c_0 = 0.025779$, $M_{b+1} = -0.081358$, $M_{b+2} = 0.224798$, respectively.

Now, we can find post-departure epoch probabilities from Equation (9). The results are presented in Table 1. We note that the results of queue-length distribution at post-departure epoch match the results given by Yi et al. [13], but the random epoch does not. We have also computed queue-length distributions at random epochs using their method for checking purposes. They match perfectly with our results. However, the results presented in the paper by [13] are different. Thus, various performance measures are also not the same.

Table 1. Queue-length distributions at various epochs for the $Geo/NB_2^{2,Y}/1/10$ queue.

| | P_j^+ | | | | $\{P_{j,0}, P_{j,1}\}$ | | |
	$\rho = 0.2$	$\rho = 0.5$	$\rho = 0.8$		$\rho = 0.2$	$\rho = 0.5$	$\rho = 0.8$
P_0^+	0.499734	0.157755	0.030728	$P_{0,0}$	0.244991	0.066960	0.010330
P_1^+	0.340335	0.290459	0.113199	$P_{1,0}$	0.411838	0.190246	0.048385
P_2^+	0.118658	0.246052	0.169892	$P_{0,1}$	0.239989	0.301616	0.170317
P_3^+	0.031582	0.148137	0.161190	$P_{1,1}$	0.077189	0.206250	0.172979
P_4^+	0.007532	0.079398	0.135131	$P_{2,1}$	0.019965	0.116185	0.148409
P_5^+	0.001694	0.040301	0.107666	$P_{3,1}$	0.004694	0.060454	0.119818
P_6^+	0.000368	0.019932	0.084302	$P_{4,1}$	0.001047	0.030223	0.093880
P_7^+	0.000078	0.009692	0.065970	$P_{5,1}$	0.000226	0.014762	0.071726
P_8^+	0.000016	0.004598	0.049310	$P_{6,1}$	0.000048	0.007247	0.059329
P_9^+	0.000003	0.002109	0.033773	$P_{7,1}$	0.000010	0.003379	0.041064
P_{10}^+	0.000001	0.001567	0.048838	$P_{8,1}$	0.000002	0.001560	0.027772
				$P_{9,1}$	0.000000	0.000665	0.016418
				$P_{10,1}$	0.000000	0.000452	0.019571
Sum	1.000000	1.000000	1.000000	Sum	1.000000	1.000000	1.000000
				L_q	0.548733	1.095056	2.820867
				W_q	3.919521	3.130145	5.137817
				PBL	0.000000	0.000452	0.019571

Remark 9. *It may be noted that $P_N^+(z)$ is a polynomial in both cases, as can be seen in Table 1. The same applies to other cases as well.*

Example 2. *$Geo/DPH^{a,Y}/1/N$queue. The service-time distribution is assumed to be discrete phase-type (DPH) having $s_k = \alpha T^{k-1}T^0$, $k = 1, 2, \ldots$, $T^0 = e - Te$, where e is the appropriate column vector with all elements equal in size. This gives the PGF of service-time distribution as $S(z) = z\alpha(I - zT)^{-1}T^0$, $|z| \leq 1$. Table 2 shows the queue length distributions at different times employing the DPH service time distribution. For the first example of Table 2, we suppose*

$$\alpha = \begin{bmatrix} 0.40 & 0.50 & 0.10 \end{bmatrix}, \quad T = \begin{bmatrix} 0.10 & 0.20 & 0.05 \\ 0.30 & 0.15 & 0.10 \\ 0.20 & 0.50 & 0.10 \end{bmatrix},$$

with $E(S) = 2.005267$ and the other parameters are $\lambda = 0.7$, $N = 15$, $y_3 = 0.5$, $y_5 = 0.3$, $y_8 = 0.2$ with $E(Y) = 4.6$. Here the denominator of $P_N^+(z)$ has eleven distinct roots, out of which three roots are outside the unit circle, and they are $\gamma_{b+1} = 2.254374$, $\gamma_{b+2} = -9.788249$, and $\gamma_{b+3} = -222.447161$. From Equation (8),

$$P_N^+(z) = T\left(-3.63007 + \frac{1.301912}{z - 2.254374} - \frac{0.702111}{z + 9.788249} + \frac{930.333318}{z + 222.447161}\right), \tag{15}$$

where $T = -1.755933$.

In the second example of Table 2, we assume $\alpha = \begin{bmatrix} 0.60 & 0.40 \end{bmatrix}$, and $T = \begin{bmatrix} 0.5 & 1/3 \\ 1/3 & 1/3 \end{bmatrix}$ with $E(S) = 4.2$ and the other parameters are $\lambda = 0.84$, $N = 50$, $y_3 = 0.7$, $y_4 = 0.2$, $y_5 = 0.1$ with $E(Y) = 3.4$. As $\rho = 1.0376$, we have one root in the range $(0, 1)$ from the remaining four roots, and the other three are outside the circle of unity. From Equation (8),

$$P_N^+(z) = T\left(0.000674 + \frac{5.2 \times 10^{-21}}{z - 1.375407} + \frac{0.001238}{z - 0.974373} - \frac{0.006324}{z - 16.101113} - \frac{4.900051 \times 10^{-13}}{z - 16.100784}\right), \tag{16}$$

where $T = -7.569441$.

Table 2. Queue-length distributions at various epochs for the $Geo/DPH^{3,Y}/1/15$ and $Geo/DPH^{3,Y}/1/50$ queues.

	$Geo/DPH^{3,Y}/1/15$				$Geo/DPH^{3,Y}/1/50$		
	$y_3 = 0.5, y_5 = 0.3, y_8 = 0.2$				$y_3 = 0.7, y_4 = 0.2, y_5 = 0.1$		
	$\lambda = 0.7, \rho = 0.30515$				$\lambda = 0.84, \rho = 1.0376$		
j	P_j^+	$P_{j,0}$	$P_{j,1}$	j	P_j^+	$P_{j,0}$	$P_{j,1}$
0	0.170394	0.055766	0.258523	0	0.001537	0.000432	0.009680
1	0.469965	0.209576	0.111943	1	0.009682	0.00315	0.009988
2	0.200698	0.27526	0.049464	2	0.010114	0.005989	0.010254
3	0.088375		0.021963	3	0.010392		0.010524
4	0.039275		0.00974	4	0.010666		0.010801
5	0.017414		0.004321	5	0.010946		0.011085
10	0.000299		0.000074	10	0.012463		0.012621
11	0.000133		0.000032	20	0.016158		0.016363
12	0.000059		0.000014	30	0.020947		0.021213
13	0.000026		0.000005	40	0.027156		0.027501
14	0.000012		0.000001	49	0.034304		0.009883
15	0.000005		0.000004	50	0.035206		0.049366
Sum	1.000000	0.540603	0.459397	Sum	1.000000	0.009570	0.990430
	$L_q = 1.1208, W_q = 1.6012$				$L_q = 29.6650, W_q = 7.1493$		
	$PBL = 0.000004$				$PBL = 0.049366$		

Remark 10. *It may be noted that Equations (15) and (16) are polynomials since the coefficients of z^b are zero. This may be seen in Table 1. This also applies to all the examples that follow.*

Example 3. *As before, in this example, we consider two cases: (i) $Geo/MGeo_2^{2,Y}/1/10$ queue. Here, the service time is a mixture of two geometric distributions with PGF $S(z) = \varsigma_1 \frac{\mu_1 z}{1-(1-\mu_1)z} + \varsigma_2 \frac{\mu_2 z}{1-(1-\mu_2)z}$, where $\varsigma_1 + \varsigma_2 = 1$. The parameters are taken as $E(S) = 2.83333$, $N = 10$, $\varsigma_1 = 0.6$, $\mu_1 = 0.4$, $\mu_2 = 0.3$, $y_2 = 0.2$, $y_3 = 0.1$, $y_4 = 0.7$, $E(Y) = 3.5$, $\lambda = 0.14$. The denominator of $P_N^+(z)$ has six distinct roots, out of which two roots are outside the unit circle, and they are $\gamma_{b+1} = 4.031316$ and $\gamma_{b+2} = 5.727647$. From Equation (8),*

$$P_N^+(z) = T\left(-0.561775 - \frac{1.819848}{z - 4.031316} - \frac{4.545280}{z - 5.727647}\right),$$

where $T = 1.0000001$.

(ii) $Geo/D^{2,Y}/1/50$ queue. Here, we consider service-time distribution as deterministic with PGF $S(z) = z^k$ for some constant $k = 4, s_k = 1$. The parameters are taken as $E(S) = 4$, $N = 50$, $y_2 = 0.5$, $y_3 = 0.2$, $y_4 = 0.1$, $y_5 = 0.1$, $y_6 = 0.1$, $E(Y) = 3.1$, and $\lambda = 0.4$. In this case, the denominator of $P_N^+(z)$ has eight distinct roots, out of which two roots are outside the unit circle, and they are $\gamma_{b+1} = 5.019701$ and $\gamma_{b+2} = -11.795711$. From Equation (8),

$$P_N^+(z) = T\left(-1.316734z^2 + 1.021799z - 102.664691 + \frac{879.864750}{z + 11.795711} - \frac{141.482007}{z - 5.019701}\right),$$

where $T = 1$. Table 3 presents queue-length distributions at various epochs when the service-time distributions are a mixture of geometric and deterministic.

Table 3. Queue-length distributions at various epochs for the $Geo/MGeo_2^{2,Y}/1/10$ and $Geo/D^{2,Y}/1/50$ queues.

$Geo/MGeo_2^{2,Y}/1/10$ $y_2 = 0.2, y_3 = 0.1, y_4 = 0.7$ $\mu_1 = 0.4, \mu_2 = 0.3, \varsigma_1 = 0.6$ $\lambda = 0.14, \rho = 0.11333$				$Geo/D^{2,Y}/1/50$ $y_2 = 0.5, y_3 = 0.2, y_4 = 0.1$ $y_5 = 0.1, y_6 = 0.1$ $\lambda = 0.4, \rho = 0.51613$			
j	P_j^+	$P_{j,0}$	$P_{j,1}$	j	P_j^+	$P_{j,0}$	$P_{j,1}$
0	0.683222	0.339297	0.15549	0	0.112580	0.052650	0.353602
1	0.250531	0.463714	0.032491	1	0.313097	0.199077	0.253667
2	0.051967		0.006998	2	0.337945		0.107802
3	0.011114		0.001551	3	0.177390		0.027021
4	0.002447		0.000352	4	0.048246		0.004913
5	0.000553		0.000082	5	0.008517		0.001019
6	0.000128		0.000019	10	0.000003		0.000003
7	0.000030		0.000005	20	0.000000		0.000000
8	0.000007		0.000001	30	0.000000		0.000000
9	0.000002		0.000000	40	0.000000		0.000000
10	0.000001		0.000000	50	0.000000		0.000000
Sum	1.000000	0.803010	0.196990	Sum	1.000000	0.251727	0.748273

$$L_q = 0.516831, W_q = 3.691651 \qquad L_q = 0.775713, W_q = 1.939284$$
$$PBL = 0.000000 \qquad PBL = 0.000000$$

Example 4. *Here, we consider two cases. Table 4 presents the results of $Geo/G^{a,Y}/1/\infty$ which can be found from the $Geo/G^{a,Y}/1/N$ system by assuming $\rho < 1$ and buffer capacity N appropriately large. We can easily compute the queue-length distributions of infinite queue capacity from finite queue capacity by assuming $\rho < 1$ and N sufficiently large.*
(i) $Geo/NB^{a,Y}/1/\infty$ queue. Here, we assume negative binomial (NB) service time distribution and the parameters are taken as $\lambda = 0.703$, $E(S) = 5$, $y_2 = 0.5$, $y_4 = 0.2$, $y_5 = 0.1$, $y_6 = 0.1$, $y_8 = 0.1$ with $E(Y) = 3.7$. From Equation (8), we have

$$P_N^+(z) = T\left(0.005129 + \frac{0.03006}{z - 2.520637} - \frac{0.00745}{z - 1.042657}\right),$$

where $T = 6.250024$.
(ii) $Geo/DPH^{a,Y}/1/\infty$ queue. For a DPH service time distribution, the settings are chosen as
$$\alpha = \begin{bmatrix} 0.60 & 0.40 \end{bmatrix} \text{ and } \mathbf{T} = \begin{bmatrix} 0.5 & 1/3 \\ 1/3 & 1/3 \end{bmatrix}.$$
with $E(S) = 4.2$ and the other parameters are $\lambda = 0.58$, $y_2 = 0.5$, $y_3 = 0.3$, $y_4 = 0.2$, and $E(Y) = 2.7$. From Equation (8), we have

$$P_N^+(z) = T\left(-0.001463 + \frac{0.019982}{z - 22.872839} - \frac{3.590919 \times 10^{-14}}{z - 22.870101} - \frac{0.004169}{z - 1.080550} + \frac{1.09 \times 10^{-20}}{z - 1.543692}\right),$$

where $T = 20.250$.

Table 4. Queue-length distributions at various epochs when $N \to \infty$.

j	$Geo/NB^{2,Y}/1/\infty$ $y_2 = 0.5, y_4 = 0.2, y_5 = 0.1, y_6 = 0.1$ $y_8 = 0.1, \lambda = 0.703, \rho = 0.95$			$Geo/DPH^{2,Y}/1/\infty$ $y_2 = 0.5, y_3 = 0.3, y_4 = 0.2,$ $\lambda = 0.58, \rho = 0.902$		
	P_j^+	$P_{j,0}$	$P_{j,1}$	P_j^+	$P_{j,0}$	$P_{j,1}$
0	0.002177	0.000616	0.028874	0.030809	0.011992	0.070392
1	0.013258	0.004369	0.034832	0.071536	0.039836	0.065422
2	0.029345		0.036239	0.066885		0.060557
3	0.034741		0.035880	0.061929		0.056044
4	0.035937		0.034858	0.057314		0.051866
5	0.035505		0.033609	0.053041		0.047999
10	0.029400		0.027374	0.036007		0.032584
20	0.019366		0.018028	0.016594		0.015016
30	0.012753		0.011872	0.007647		0.006920
50	0.005531		0.005149	0.001624		0.001470
100	0.000685		0.000638	0.000034		0.000031
150	0.000085		0.000079	0.000001		0.000001
200	0.000011		0.000010	0.000000		0.000000
250	0.000001		0.000001	0.000000		0.000000
300	0.000000		0.000000	0.000000		0.000000
500	0.000000		0.000000	0.000000		0.000000
Sum	1.000000	0.004986	0.995014	1.000000	0.051828	0.948172
	$L_q = 23.80989, W_q = 33.86898$			$L_q = 11.815062, W_q = 20.370796$		

In Figure 2, we compare the processing times to calculate probabilities at post-departure using the proposed technique and the method used by [13] (solving a linear system of equations (SLSE)) against finite buffer capacity. We take the NB service-time distribution in two stages with the input parameters in the same way as in Table 1 for $\rho = 0.2$. We notice that, with the increase of N, the time needed by the roots method remains almost static, whereas the method used by [13] takes more time for larger buffer size N and increases remarkably as N increases. In the roots method, the variation of the time is minimal. This is because of the initial estimation of the Newton iterative method for the calculation of polynomial roots. The application of the SLSE process is low in reliability and time-consuming. The root method provides a faster solution and superior performance in solving a linear system of method equations both in speed and reliability.

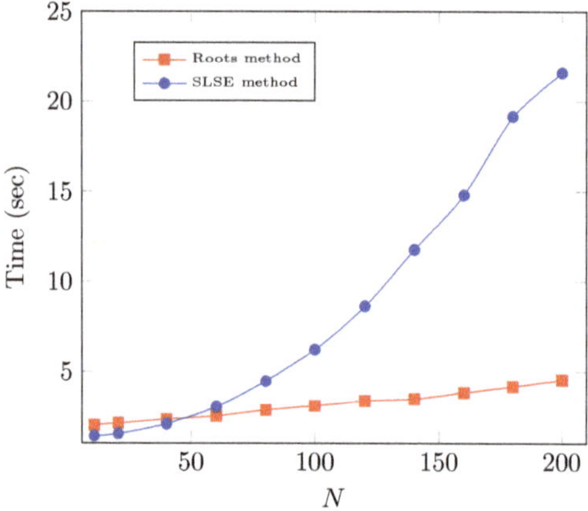

Figure 2. Time (in seconds) needed to calculate post-departure probabilities.

Figure 3 shows the roots of the characteristic equation for the number in the queue with NB service time distribution having 4 successes is the convolution of 4 geometric distributions. Here we consider the parameters as $\lambda = 0.81$, $y_4 = 0.4$, $y_6 = 0.1$, $y_{12} = 0.1$, $y_{25} = 0.2$, $y_{30} = 0.1$, $y_{36} = 0.1$, and $\rho = 0.54$. There are 40 of roots inside, on, and out of the unit circle for the assumed parameters. Here the characteristic equation is

$$z^{36}(-0.886 + 0.486\, z)^4 - 0.0256(0.19 + 0.81\, z)^4 (0.1 + 0.4\, z^{32} + 0.1\, z^{30}$$
$$+0.1\, z^{24} + 0.2\, z^{11} + 0.1\, z^6). \tag{17}$$

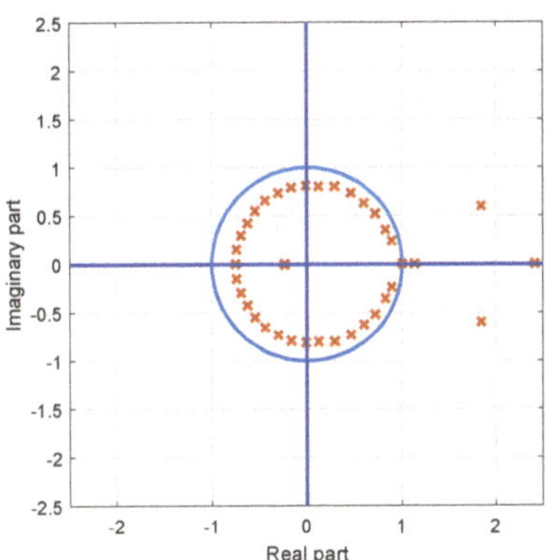

Figure 3. The 40 roots of Equation (17) when NB service-time distribution.

6. Conclusions

This article focuses on the $Geo/G^{a,Y}/1/N$ queue length distributions at various points in time. We use the roots of the associated characteristic equation to determine a unified way to compute performance measures for both infinite- and finite-buffer systems. Queue length distributions at a post-departure time are computed using an embedded Markov chain method. We obtain associations between queue length distributions at several time points by applying system state conditioning and discrete renewal theory. Several performance indices have been carried out, such as the blocking probability, the average wait time in the queue, and the average number of jobs in the queue. We illustrate them by using different numerical outcomes. The approach discussed in this paper can be used to cover cases when customers arrive in groups or even when arrivals are correlated (D-MAP-discrete Markovian arrival process).

Author Contributions: Conceptualization, M.C. and V.G.; methodology, V.G. and M.C.; Writing—original draft preparation, V.G.; supervision, M.C.; funding acquisition, M.C. All authors have read and agreed to the published version of the manuscript.

Funding: This research was funded by FMAS IO 757193 (C143).

Data Availability Statement: Not applicable.

Conflicts of Interest: The authors declare no conflict of interest.

Appendix A. The Continuous-Time Case

Here, we consider the relation between the discrete-time $Geo/G^{a,Y}/1/N$ queue and its continuous-time analogue. Let the time axis be divided into periods of uniform length Δu with $\Delta u > 0$ sufficiently small. In $Geo/G^{a,Y}/1/N$, since the inter-arrival times (u) follow geometric distribution, the arrivals will follow binomial distribution which, as we have seen earlier, leads to PGF for binomial distribution. In the continuous-time case, geometric tends to exponential, and binomial tends to Poisson distribution, and the PGF tends to

Laplace transform. Let us discuss this analytically. Assume that the inter-arrival times in the case of $M/G^{a,Y}/1/N$ have a rate α. Then, $\lambda = \alpha\Delta u + o(\Delta u)$. In the discrete-case, let the service times S be in multiples of Δu with probability $P(S = \ell\Delta u) = s_\ell$ and $\sum_{\ell=1}^{\infty} s_\ell = 1$. Further, let $n\Delta u = v_n$, where the interval $[0, v_n]$ is divided into n intervals of length Δu. The PGF of an arrival (or no arrival) in the interval $(v_\ell, v_{\ell+1})$ is $(1 - \lambda + \lambda z)$. If we denote the probability density function of service times by $h(\cdot)$, then

$$P(\text{service finishes in } (u, u + \Delta u)|\text{ service time} > u) = h(u)\Delta u + o(\Delta u)$$

$$\text{and} \quad s_\ell = h(v_\ell)\Delta u + o(\Delta u).$$

When $\Delta \to 0$, the PGF $S(1 - \lambda + \lambda z)$ changes to a Laplace transform. Using the definition of Lebesgue integration and taking the limit as $\Delta \to 0$ and $\lambda = \alpha\Delta$, we have

$$\lim_{\Delta \to 0} K(z) = \lim_{\Delta \to 0} \sum_{\ell=1}^{\infty} s_\ell (1 - \lambda + \lambda z)^\ell = \bar{h}(\alpha - \alpha z).$$

The proof of the above is not discussed in detail here since the method applied can be found in [28]. Now, using $\lambda = \alpha\Delta$, $K(z) = \bar{h}(\alpha - \alpha z)$ in (5) and taking the limit as $\Delta \to 0$, we have

$$P_N^+(z) = \frac{\bar{h}(\alpha - \alpha z)\left[\sum_{i=0}^{a-1} P_i^+\left(z^b - z^i Y'(z)\right) + \sum_{i=a}^{b-1} P_i^+\left(\sum_{r=i}^{b} y_r z^b - z^i \sum_{r=i}^{b} y_r z^{b-r}\right)\right]}{z^b - \bar{h}(\alpha - \alpha z)Y'(z)},$$

the connections for $M/G^{a,Y}/1/N$ system. Taking $a = 1$, we have

$$P_N^+(z) = \frac{\bar{h}(\alpha - \alpha z)\left[\sum_{i=0}^{b-1} P_i^+ z^b\left(\sum_{r=i}^{b} y_r - z^i \sum_{r=i}^{b} y_r z^{-r}\right)\right]}{z^b - \bar{h}(\alpha - \alpha z)Y'(z)},$$

which matcheswith [29].

Note that we cannot obtain results of discrete-time analogue from [29], that is, the converse is not true.

References

1. Chaudhry, M.L.; Templeton, J.G.C. *First Course in Bulk Queues*; Wiley: New York, NY, USA, 1983.
2. Medhi, J. *Stochastic Models in Queueing Theory*; Elsevier: Amsterdam, The Netherlands, 2002.
3. Bruneel, H.; Kim, B.G. *Discrete-Time Models for Communication Systems Including ATM*; Springer Science & Business Media: New York, NY, USA, 2012; Volume 205.
4. Hunter, J.J. *Mathematical Techniques of Applied Probability: Discrete Time Models: Basic Theory*; Academic Press: New York, NY, USA, 2014; Volume 1.
5. Takagi, H. *Queueing Analysis: Discrete-Time Systems*; North Holland: Amsterdam, The Netherlands, 1993; Volume 3.
6. Gravey, A.; Hébuterne, G. Simultaneity in discrete-time single server queues with bernoulli inputs. *Perform. Eval.* **1992**, *14*, 123–131. [CrossRef]
7. Chaudhry, M.L.; Chang, S.H. Analysis of the discrete-time bulk-service queue Geo/G^Y/1/N+ B. *Oper. Res. Lett.* **2004**, *32*, 355–363. [CrossRef]
8. Denteneer, D.; Janssen, A.J.; Van Leeuwaarden, J. Moment inequalities for the discrete-time bulk service queue. *Math. Methods Oper. Res.* **2005**, *61*, 85–108. [CrossRef]
9. Goswami, V.; Mohanty, J.; Samanta, S.K. Discrete-time bulk-service queues with accessible and non-accessible batches. *Appl. Math. Comput.* **2006**, *182*, 898–906. [CrossRef]
10. Gupta, U.C.; Goswami, V. Performance analysis of finite buffer discrete-time queue with bulk service. *Comput. Oper. Res.* **2002**, *29*, 1331–1341. [CrossRef]
11. Janssen, A.J.; VanLeeuwaarden, J. Analytic computation schemes for the discrete-time bulk service queue. *Queueing Syst.* **2005**, *50*, 141–163. [CrossRef]
12. Sivasamy, R.; Pukazhenthi, N. A discrete time bulk service queue with accessible batch: Geo/$NB^{(L,K)}$/1. *Opsearch* **2009**, *46*, 321–334. [CrossRef]

13. Yi, X.W.; Kim, N.K.; Yoon, B.K.; Chae, K.C. Analysis of the queue-length distribution for the discrete-time batch-service Geo/$G^{a,Y}$/1/K queue. *Eur. J. Oper. Res.* **2007**, *181*, 787–792.

14. Banerjee, A.; Gupta, U.; Goswami, V. Analysis of finite-buffer discrete-time batch-service queue with batch-size-dependent service. *Comput. Ind. Eng.* **2014**, *75*, 121–128.

15. Zeng, Y.; Xia, C.H. Optimal bulking threshold of batch service queues. *J. Appl. Probab.* **2017**, *54*, 409–423.

16. Chaudhry, M.L.; Harris, C.M.; Marchal, W.G. Robustness of rootfinding in single-server queueing models. *ORSA J. Comput.* **1990**, *2*, 273–286. [CrossRef]

17. Gouweleeuw, F.N. A General Approach to Computing Loss Probabilities in Finite Buffer Queues. Ph.D. Thesis, Vrije Universiteit Amsterdam, Amsterdam, The Netherlands, 1996.

18. Powell, W.B. Stochastic Delays in Transportation Terminals: New Results in the Theory and Application of Bulk Queues. Ph.D. Thesis, Massachusetts Institute of Technology, Cambridge, MA, USA, 1981.

19. Kendall, D.G. Some recent work and further problems in the theory of queues. *Theory Probab. Its Appl.* **1964**, *9*, 1–13. [CrossRef]

20. Kleinrock, L. Theory. In *Queueing Systems*; Wiley-Interscience: New York, NY, USA, 1975; Volume 1.

21. Neuts, M.F. *Matrix-Geometric Solutions in Stochastic Models: An Algorithmic Approach*; Courier Corporation: Washington, DC, USA, 1994.

22. Stidham, S., Jr. *Applied Probability in Operations Research: A Retrospective*; University of North Carolina, Department of Operations Research: Chapel Hill, NC, USA, 2001.

23. Chaudhry, M.L.; Goswami, V. The queue Geo/G/1/N+1 revisited. *Methodol. Comput. Appl. Probab.* **2019**, *21*, 155–168. [CrossRef]

24. Botta, R.F.; Harris, C.M.; Marchal, W.G. Characterizations of generalized hyperexponential distribution functions. *Stoch. Model.* **1987**, *3*, 115–148. [CrossRef]

25. Kobayashi, H.; Mark, B.L.; Turin, W. *Probability, Random Processes, and Statistical Analysis: Applications to Communications, Signal Processing, Queueing Theory and Mathematical Finance*; Cambridge University Press: Cambridge, UK, 2011.

26. Cohen, J.W. *The Single Server Queue*; North Holland Publishing Company: Amsterdam, The Netherlands, 1982.

27. Boxma, O.J.; Groenendijk, W.P. Waiting times in discrete-time cyclic-service systems. *IEEE Trans. Commun.* **1988**, *36*, 164–170. [CrossRef]

28. Yang, T.; Li, H. On the steady-state queue size distribution of the discrete-time Geo/G/1 queue with repeated customers. *Queueing Syst.* **1995**, *21*, 199–215. [CrossRef]

29. Singh, V. Finite waiting space bulk service system. *J. Eng. Math.* **1971**, *5*, 241–248. [CrossRef]

mathematics

Article

Really Ageing Systems Undergoing a Discrete Maintenance Optimization

Radim Briš * and Pavel Jahoda

Department of Applied Mathematics, Faculty of Electrical Engineering and Computer Science, VSB—Technical University of Ostrava, 708 00 Ostrava-Poruba, Czech Republic
* Correspondence: radim.bris@vsb.cz

Abstract: In general, a complex system is composed of different components that are usually subject to a maintenance policy. We take into account systems containing components that are under both preventive and corrective maintenance. Preventive maintenance is considered as a failure-based preventive maintenance model, where full renewal is realized after the occurrence of every nth failure. It offers an imperfect corrective maintenance model, where each repair deteriorates the component or system lifetime, the probability distribution of which gradually changes via increasing failure rates. The reliability mathematics for unavailability quantification is demonstrated in the paper. The renewal process model, involving failure-based preventive maintenance, arises from the new corresponding renewal cycle, which is designated a real ageing process. Imperfect corrective maintenance results in an unwanted rise in the unavailability function, which can be rectified by a properly selected failure-based preventive maintenance policy; i.e., replacement of a properly selected component respecting both cost and unavailability after the occurrence of the nth failure. The number n is considered a decision variable, whereas cost is an objective function in the optimization process. The paper describes a new method for finding an optimal failure-based preventive maintenance policy for a system respecting a given reliability constraint. The decision variable n is optimally selected for each component from a set of possible realistic maintenance modes. We focus on the discrete maintenance model, where each component is realized in one or several maintenance mode(s). The fixed value of the decision variable determines a single maintenance mode, as well as the cost of the mode. The optimization process for a system is demanding in terms of computing time because, if the system contains k components, all having three maintenance modes, we need to evaluate 3^k maintenance configurations. The discrete maintenance optimization is shown with two systems adopted from the literature.

Keywords: unavailability; imperfect repair; failure-based replacement; renewal theory; optimization

MSC: 60K10; 90B25

Citation: Briš, R.; Jahoda, P. Really Ageing Systems Undergoing a Discrete Maintenance Optimization. *Mathematics* **2022**, *10*, 2865. https://doi.org/10.3390/math10162865

Academic Editors: Gurami Tsitsiashvili and Alexander Bochkov

Received: 15 July 2022
Accepted: 9 August 2022
Published: 11 August 2022

Publisher's Note: MDPI stays neutral with regard to jurisdictional claims in published maps and institutional affiliations.

1. Introduction

Various maintenance strategies have recently been intensively studied and developed to improve the reliability, availability and usability of relevant industrial systems. Unexpected failures may cause dangerous situations for human lives, unplanned production outages, etc. This is why systems must be protected against them. System reliability can be significantly improved by applying the optimal maintenance strategy. One can read in the work of [1] that maintenance is no longer a necessary evil and that production companies should invest in maintenance to maximize revenues. To find the best and most suitable strategy for every unit or subsystem, preferred economic decisions have to be made that make it possible to achieve a profit. Therefore, evaluation of the performances of different strategies is often used ([1,2]).

We can distinguish between the different maintenance strategies that can be used: corrective maintenance (CM) and preventive maintenance (PM) strategies. CM strategies, often known as restoration or repair strategies, are only launched when a failure occurs and the system is broken. The system is then returned to a functioning state by applying a maintenance action. PM strategies aim to prevent the system from undergoing undesired breakdowns. PM is usually carried out on operating systems and it reduces ageing processes, which means that the probability of system failures is decreased. Maintenance modelling is a dynamically developing recent scientific discipline, and we do not aim to present a general overview of references on maintenance here. However, several papers can be discussed, some of them dealing with CM, others with PM. For example, authors of [3] mention that there are several different types of PM and CM actions depending on the degree of restoration, including perfect maintenance action, which restores the system to an as-good-as-new function state; and imperfect maintenance action, which can have several restoration levels between perfect maintenance and minimal maintenance, with the latter restoring the system to the same state it was in just before the failure occurred, with the same failure rate that it had just before the minimal maintenance action.

A repair action can gradually restore a system to its initial level, with the system being returned step-by-step to the operating state, resulting in perfect CM. This approach is in detail introduced for example in [4] and the authors call it a gradual CM repair strategy. In general, we can say that imperfect maintenance is a kind of intervention where the system is returned to someplace between being as good as new and as bad as it used to be. However, for any maintenance intervention, it is necessary to establish the level of imperfect maintenance. Authors in [5] introduced such imperfect restoration when they assumed that the system age is affected by the kind of maintenance intervention. Apart from the system age, the failure rate may also become worse due to maintenance actions ([6]). Both age shortening and failure rate adaptation are assumed in the so-called hybrid model, which seems to be a more realistic model (see more in [7,8]). Whereas the former authors addressed the optimal maintenance decision for binary systems to maximize the reliability of the next mission under imperfect maintenance, the latter introduced the imperfect maintenance model, where fixed maintenance action corrects a system functionality to any state between minimal repair and full renewal; i.e., perfect repair. For extensive discussions regarding imperfect maintenance models, readers may refer to authors in [9,10].

In this article, we use the imperfect CM process, which degrades the system lifetime at any CM intervention due to the growing failure rate. Gradual changes in the failure rate result in corresponding changes in the lifetime probability distribution. The basic aim of this article is to describe a realistic failure-based PM model that undergoes a process of discrete optimization. PM is considered a failure-based preventive maintenance model (FBM), where full renewal is realized at the occurrence of every nth failure. This model is related to the imperfect CM process, an overview of which is provided in [11]. For instance, a comparable strategy, where a unit is replaced at the nth failure and $(n-1)$ previous failures are repaired with minimal repair, was proposed in the cost-focused study [12]. However, the author supposed that the failure rate would not be violated after carrying out minimal repair. Stochastic models that describe the failure pattern of repairable units subject to minimal maintenance are discussed in [13]. The unit can be replaced at time T or at the nth failure, whichever occurs first, and n can be minimized in the context with both repair and replacement costs, as is developed in [14]. We use the imperfect CM model, where each repair action deteriorates the component or system lifetime, the probability distribution of which is gradually changed via the increasing failure rate. This model is of particular interest for experts analyzing components of power distribution networks ([15]), the reliability of which will be studied by the authors of this article in the near future.

Innovative reliability mathematics for the unavailability quantification of components is presented in this article. The new renewal process model, involving the FBM, is designated as a real ageing process. The imperfect CM results in an unwanted rise in the unavailability function, which can be rectified with a properly selected FBM process. This means that the renewal of a component starts with the occurrence of the nth failure. The number n is considered a decision variable, whereas cost is an objective function in the optimization process.

This article describes a new method that can be used to find an optimal FBM strategy to solve a particular optimization problem while respecting a given reliability constraint. The above-mentioned decision variable determining the different maintenance modes of a system component is optimally selected from a set of possible realistic maintenance modes. Thus, the discrete maintenance model is considered, where each component can work in one or several maintenance modes. The fixed value of the decision variable n determines one maintenance mode of the component, which predetermines both the unavailability course and cost. Different maintenance modes of system components result in different system configurations, each having a specific unavailability course, as well as cost. The optimization process is demanding in terms of computing time because a complex system can have many maintenance configurations. The discrete maintenance optimization is demonstrated with two systems adopted from the literature.

2. Optimization Problem

Each optimization problem works on the presumption that an objective function $f(\mathbf{x})$ that varies in a given range must be optimized; i.e., either maximized or minimized, constrained by several restrictions imposed on the decision variables. The optimization problem in this article can be formulated using the following objective function $f(\mathbf{x})$, which represents the minimum cost:

$$f(\mathbf{x}) = \min C_S \tag{1}$$

$$\text{subject to the constraint } U_S(\mathbf{x}) \leq U_0 \tag{2}$$

where $\mathbf{x} = (x_1, \dots, x_k) \in \mathbf{R}^k$ is a decision variable, and k is the number of system components each having the decision variable $x_i = n_i$, which can be optimized if needed. Each component undergoes a real ageing process, including imperfect CM, until the occurrence of the n_i-th failure, which starts its restoration (renewal).

In most cases, both $f(\mathbf{x})$ and $U_S(\mathbf{x})$ are complicated, either linear or nonlinear, functions of the decision variable vector:

$$\mathbf{x} = (x_1, \dots, x_k) = (n_1, \dots n_k)$$

that constitute parameters for which optimal values must be found.

3. Discrete Maintenance Model

Maintenance optimization can be classified in different ways. A recent thorough classification can be found in [16]. Concerning the optimization outcome and decision variables, the discrete maintenance model can be classified as a process that finds optimized parameter values defining a single maintenance strategy selected a priori; e.g., in this paper, the type of action performed (repair, replacement). There are different optimization approaches that take into account the previously selected decision variables. The methodology used in this paper can be included among the mathematical approaches in which the optimization problem is formulated utilizing mathematical equations, which are then solved using differential calculus to identify optimal parameters for the maintenance strategy.

We introduced a discrete maintenance model for real multi-component systems with non-identical components for the first time in [17], in which systems with repairable components and latent failures were optimized using changeable periods of inspection as a decision variable. Real complex systems are composed of a finite number of repairable and maintained components. Each component can be operated in different discrete maintenance modes. A maintenance mode of the i-th component is determined by a prescribed value of the decision variable ($x_i = n_i$) that influences the maintenance cost of the mode. Given that a system is composed of k components and each component has four maintenance modes in general, we have to investigate 4^k maintenance configurations of the system. Any system configuration can be described by a maximal system unavailability $U_S(\mathbf{x})$ and a total cost

C_S, which is usually obtained as a sum of the costs of all component modes forming the configuration. The optimal system configuration is detected under requirements (1) and (2). This maintenance model is defined in this article as a discrete maintenance model.

Similar models of maintenance optimization have been used in other publications. For example, the work of authors in [18–20] addresses the optimization problem under maintenance policies including two decision variables. One of them is the maximum number of failures before the system undergoes a perfect restoration and the second is the inspection interval used to detect hidden failures. In [21] authors consider only one component system subject to two types of age-dependent failure. Catastrophic failures are detected through periodic inspections, whereas minor failures are followed by minor repairs. The system is preventively replaced either at an optimal multiple of the inspection time or after the n-th minor failure, whichever comes first. Both parameters are decision variables. The cost per unit of time is an objective function. The objective is to obtain a cost-minimizing policy. Authors in [22] also present a maintenance model for a system subject to two types of unrevealed failures: minor and catastrophic. The system is replaced at the occurrence of the n-th minor failure, a catastrophic failure, or due to working age, etc.

4. Unavailability Analysis and Cost Model

4.1. The Method for Unavailability Calculation for a Complex System

The basic methodology, including algorithms, for unavailability calculation for a complex system with maintenance was developed in [23,24]. The system structure and its functionality are described through the use of a directed acyclic graph (AG). An AG contains nodes and edges. The system function or non-function state constitutes the highest node, which is at the top of the AG. Subsystems (components) are described through internal (terminal) nodes, all of which are interconnected by edges. An AG cannot contain feedback loops because it is acyclic. Terminal nodes characterize the stochastic behavior of the input components of a system, which means that each terminal node must be provided with information about the probability distribution of its lifetime, as well as maintenance characteristics and parameters. Stemming from this information, the unavailability function of each terminal node is further computed through the renewal process model described in the following section. The final system unavailability function is obtained through the unavailability functions of all terminal nodes. Other details concerning the computing algorithm that calculates the system unavailability function from component (i.e., terminal node) unavailability functions are described in our previous research work in ([17], pp. 86–87).

4.2. Unavailability Analysis of Terminal Nodes

The reliability mathematics used in this section results partially from the work of authors in [25] and partially from other articles presented in the references ([26–30]). The mathematics used in these sources was developed to a large extent to consider the new renewal cycle of a terminal node, as introduced below.

A complex maintained system consists of particular components that, in the context of the AG system structure, are denoted terminal nodes. In this section, the unavailability function of a terminal node is investigated. The renewal cycle of a terminal node starts at time $t = 0$. The first failure occurs at the time X_1 and CM of the node starts immediately. This CM action takes the time Y_1. The next failure occurs after the time X_2 after the end of the repair of the first failure elapses. In this way, the renewal cycle continues until the nth failure occurs. This is followed by the FBM replacement of the node. When this is completed, the renewal cycle ends, as is demonstrated in Figure 1, where the length of the renewal cycle is T. We will call the progress over one renewal cycle a real ageing process. This evolution is characterized by gradually changing probability distributions of random variables $X_1, X_2, \ldots X_n$ due to different degradation processes caused by CM.

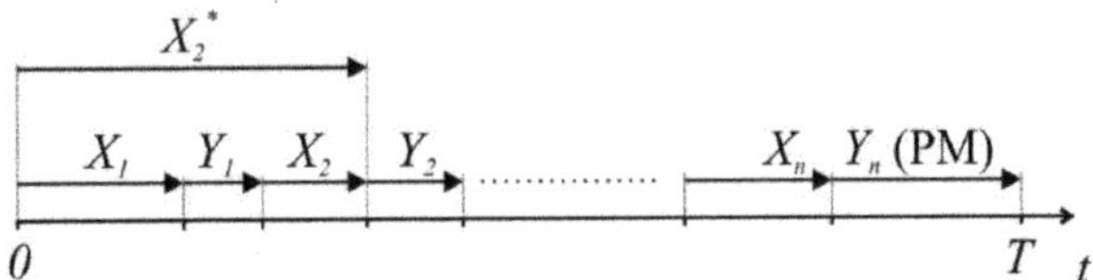

Figure 1. The first renewal cycle. The time from the beginning of the renewal cycle to the occurrence of the second failure is a random variable $X_2^* = X_1 + Y_1 + X_2$, etc. The renewal cycle T is terminated by the n-th failure and followed by the replacement. The replacement time is the random variable Y_n.

First, the unavailability function of the terminal node at a given time t in the first renewal cycle is determined.

The random variables $X_1, \ldots, X_n, Y_1, \ldots, Y_n$ are supposed to be independent.

The terminal node is out of service at time t only if it is in the process of being repaired (or replaced) after the i-th failure. This happens only when $0 \leq X_i^* \leq t \leq X_i^* + Y_i$ for some $i \in \{1, \ldots, n\}$.

It is well-known that

$$P(X_i^* + Y_i < t) = F_{X_i^* + Y_i}(t) = \int_{-\infty}^{\infty} F_{Y_i}(t - x)\mathrm{d}F_{X_i^*}(x)$$

Hence

$$P(t \leq X_i^* + Y_i) = \overline{F}_{X_i^* + Y_i}(t)$$

$$= 1 - \int_{-\infty}^{\infty} F_{Y_i}(t - x)\mathrm{d}F_{X_i^*}(x)$$

$$= \int_{-\infty}^{\infty} \big(1 - F_{Y_i}(t - x)\big)\mathrm{d}F_{X_i^*}(x)$$

$$= \int_{-\infty}^{\infty} \overline{F}_{Y_i}(t - x)\mathrm{d}F_{X_i^*}(x)$$

Let us denote the probability that the terminal node is unavailable at time t due to its CM or FBM after the i-th failure $P_i(t)$. It follows that

$$P_i(t) = P(0 \leq X_i^* \leq t \leq X_i^* + Y_i) = \int_0^t \overline{F}_{Y_i}(t - x)\mathrm{d}F_{X_i^*}(x)$$

The probability that the node is out of order at time t in the first renewal cycle (let us denote it $u_1(t)$) fulfils the following equation:

$$u_1(t) = \sum_{i=1}^n P_i(t) = \sum_{i=1}^n \int_0^t \overline{F}_{Y_i}(t - x)\mathrm{d}F_{X_i^*}(x) \tag{3}$$

The length of the k-th renewal cycle is the random variable T_k and the k-th renewal cycle is the time interval (s_{k-1}, s_k) where $s_0 = 0$ and s_k is the value of the random variable $S_k = \sum_{i=1}^k T_i$, as is shown in Figure 2.

Figure 2. Assuming that the FBM restores the system to an as-good-as-new state, it holds that all renewal cycles are equivalent; i.e., $T_i = T_j = T$ for all $i, j \in \mathbb{N}$.

The unavailability function of a terminal node describes the probability that the node is out of order at a given time t. It can be determined as follows. Let us denote this probability $u(t)$. It is obvious that

$$u(t) = \sum_{k=1}^{\infty} u_k(t), \tag{4}$$

where $u_k(t)$ is the probability that the node is out of order at time t, which belongs to the k-th renewal cycle $\langle s_{k-1}, s_k \rangle$.

The value $u_1(t)$ is given by Formula (3). The values $u_k(t)$ for $k = 2, 3, \ldots$ can be determined in the following way, assuming that the FBM restores the terminal node to an as-good-as-new state. It holds that $u_2(t) = u_1(t - s_1)$, $u_3(t) = u_1(t - s_2), \ldots$ and, in general, for $k \geq 2$ (see Figure 3):

$$u_k(t) = u_1(t - s_{k-1}) = \int_0^t u_1(t - s) \mathrm{d}F_{S_{k-1}}(s) \tag{5}$$

where $F_{S_{k-1}}$ is the cumulative distribution function of the random variable $S_{k-1} = \sum_{i=1}^{k-1} T_i$. Under the assumption that the FBM restores the node to an as-good-as-new state, it also holds that $T_i = T_j = T$ for all $i, j \in N$. This means that $F_{S_{k-1}}$ is the $(k-1)$-fold convolution of the cumulative distribution function F_T, where the random variable $T = \sum_{i=1}^{n} X_i + \sum_{i=1}^{n} Y_i$ is the length of the renewal cycle. Hence

$$ET = \sum_{i=1}^{n} EX_i + \sum_{i=1}^{n} EY_i$$

and F_T is the convolution of cumulative distribution functions $F_{X_1}, \ldots, F_{X_n}, F_{Y_1}, \ldots$ and F_{Y_n}.

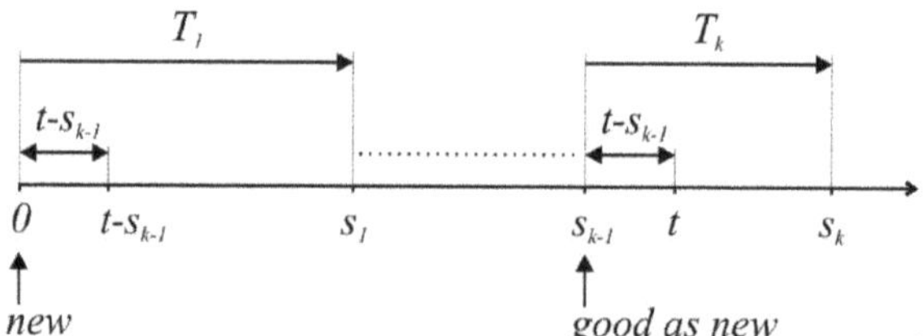

Figure 3. The value of the unavailability function $u_k(t)$ in the k-th renewal cycle for $t \in \langle s_{k-1}, s_k \rangle$ is the same as the value of the unavailability function u_1 at the time $t - s_{k-1}$.

Equations (4) and (5) then imply:

$$u(t) = u_1(t) + \sum_{k=2}^{\infty} \int_0^t u_1(t - s) \mathrm{d}F_{S_{k-1}}(s)$$

$$= u_1(t) + \sum_{k=1}^{\infty} \int_0^t u_1(t - s) \mathrm{d}F_{S_k}(s)$$

$$= u_1(t) + \int_0^t u_1(t - s) \mathrm{d}G(s), \tag{6}$$

where $G(s) = \sum_{k=1}^{\infty} F_{S_k}(s)$. Since $S_1 = T$, it holds that

$$G(s) = F_T(s) + \sum_{k=2}^{\infty} P(S_k \leq s)$$

$$= F_T(s) + \sum_{k=1}^{\infty} P(S_k + T \leq s)$$

$$= F_T(s) + \sum_{k=1}^{\infty} F_{S_k}(s - T)$$

$$= F_T(s) + \int_0^s \sum_{k=1}^{\infty} F_{S_k}(s - T)\mathrm{d}F_T(t)$$

$$= F_T(s) + \int_0^s G(s - T)\mathrm{d}F_T(t).$$

Hence, the function $G(s)$ is the solution of the integral equation

$$G(s) = F_T(s) + \int_0^s G(s - T)\mathrm{d}F_T(t).$$

Thus, knowing $G(s)$, the required unavailability function $u(t)$ can be determined as a solution of the integral Equation (6).

The starting point for the development of this methodology lies in our previous work introduced in [31], where dormant systems under inspection were investigated. The main difference is in the renewal cycle. A dormant system has a different renewal cycle because failures are not detected immediately but only at special inspection times.

4.3. Cost Model

The cost model of a system configuration can be derived by adding up all the contributions arising from both the repair and replacement processes of a mode over all of the system components. A maintenance mode of a component has two main cost contributions: the cost of FBM, given by the replacements of the component during a mission time T_M that depends on the decision variable of the component; and the cost of the imperfect repair process, which further depends on the mean number of failures during the mission time T_M and the CM parameters. In practical situations, the cost contributions result from a database for the year and give an average yearly cost for the system configurations in a monitored period. In the remainder of this article, the cost will be computed in non-identified cost units based on the summation principle.

To obtain the cost of one system configuration, we simply add up the costs of all maintenance modes of all system components. The mean cost of one maintenance mode of the j-th component $C_{T_M}(j)$ can be computed as follows:

$$C_{T_M}(j) = \left[\frac{n_R(j)}{n_j}\right] \cdot C_R(j) + \left(n_R(j) - \left[\frac{n_R(j)}{n_j}\right]\right) \cdot C_{CM}(j) \tag{7}$$

where

$$n_R(j) = \frac{T_M}{MTTF(j) + MTTR(j)}$$

$n_R(j)$ is the mean number of failures of the j-th component per mission time T_M
$MTTF(j)\ldots$ mean time to failure of the j-th component:

$$MTTF(j) = \frac{\sum_{k=1}^{n_j} MTTF_k(j)}{n_j} \tag{8}$$

$MTTF_k(j)\ldots$ mean time to the k-th failure of the j-th component
$MTTR(j)\ldots$ mean repair time of the j-th component
$n_j\ldots$ decision variable of the j-th component determining the FBM strategy
$[x]\ldots$ integral part of the real number x (i.e., $f(x) = [x]$ is the floor function)
$\left[\frac{n_R(j)}{n_j}\right]\ldots$ number of FBM replacements of the j-th component per mission time T_M
$\left(n_R(j) - \left[\frac{n_R(j)}{n_j}\right]\right)\ldots$ mean number of repairs (CM) of the j-th component
$C_R(j)\ldots$ replacement cost = cost of one FBM action for the j-th component in cost units

$C_{CM}(j)\dots$ CM cost = cost of one repair action for the j-th component

$C_{T_M}(j)\dots$ mean cost of one maintenance mode of the j-th component

It is worth noting that the mean maintenance cost of the j-th component $C_{T_M}(j)$ depends on the decision variable n_j, not only directly, as follows from Formula (7), but also indirectly via $n_R(j)$ (see Formula (8) for $MTTF(j)$). The total cost of one system configuration C_S is obtained by summing up these contributions, described by Formula (7), over all of the system components k:

$$C_S = \sum_{j=1}^{k} C_{T_M}(j) \tag{9}$$

A similar principle for the computation of the total maintenance cost for the whole of a system was used in [32].

5. Unavailability Optimization of Selected Systems and Discussion

5.1. One Maintained Component System Adopted from the Literature

First, we take into account one component system that is under both PM and CM. PM is considered as FBM, where full renewal is realized at the occurrence of every nth failure. The imperfect CM model causes a real ageing process, where each CM intervention deteriorates the system lifetime at an increasing failure rate. Proceeding from the previously derived reliability mathematics, the system unavailability function can be determined from Formula (6). The imperfect CM results in an unwanted rise in the unavailability function that can be reduced by the properly selected FBM process, meaning that renewal of the system starts with the occurrence of the n-th failure. After the renewal, the system is restarted to an as-good-as-new state. The number n is considered here as a changing decision variable permitting optimization. The exact specification of the decision variable determines a system configuration that is connected to a specific cost computed according to Formula (9), where $k = 1$.

The discrete maintenance optimization method is shown for a system adopted from [25] where a stochastic alternating renewal process model is derived but a different renewal cycle is considered. The distribution function of the first failure time X_1 is a Weibull distribution with the shape parameter $\beta = 2$ and scale parameter $\alpha = 600$ days. Imperfect CM is characterized by a random repair time with rectangular distribution in an interval of <12, 16> days, so that the mean time for CM is 14 days. We further suppose that the replacement time in the FBM model is deterministic—the renewal duration is 7 days; i.e., it is shorter than the CM time since it may be scheduled beforehand, which is given by the fact that the repair team has at its disposal information about $(n-1)$-th failure.

The real ageing process can be realized in the following way. If a failure occurs, it is followed by a standard CM action, which recovers the health of the system to some extent, but its failure rate is always worse when compared with the system health before failure. We presume that, following the first system failure and the CM action, the growth of the failure rate can be estimated by the quotient q_a, which ranges from 1 to 1.5 and by which the failure rate is multiplied. Worsening of the failure rate will continue after the second failure and will be followed by a repair time, etc., until the time of the FBM intervention; i.e., the time of the n-th failure. Provided that the first failure time X_1 follows a Weibull distribution, the failure rate can be expressed in the following way:

$$\lambda(t) = \frac{\beta}{\alpha^2}.t \tag{10}$$

X_2 also has a Weibull distribution failure rate, which is worsened as follows:

$$\lambda(t) = \frac{\beta}{\alpha^2}.q_a.t \tag{11}$$

and the worsening continues with each subsequent failure in the same way. In the time X_{n-1}, the failure rate can be expressed as:

$$\lambda(t) = \frac{\beta}{\alpha^2} \cdot q_a{}^{n-2} \cdot t \tag{12}$$

and the FBM is started at the time X_n, which launches a renewal (replacement). Thus, the system progressively deteriorates, and its lifetime distribution is modified after each CM action until an FBM time comes. This is in accordance with real practice because each failure followed by a CM action exposes the system to shocks that accumulate, and the system becomes increasingly worse. For example, the level of deterioration after the CM of circuit breakers, components of a power distribution network, has been intensively and personally discussed with experts and authors ([15]), who stated that each CM action makes the initial technical health of the component worse by approximately 10–25%. That is why we selected the following computing experiments with an appropriate value for the quotient $q_a = 1.25$, which means the limit worsening of the failure rate by 25% after each CM intervention.

The difference between real and theoretical ageing is demonstrated in Figure 4, which illustrates the time-dependent unavailability function of the system $u(t)$, computed according to Formula (6), with a mission time of 4000 days for two system modes:

1. A real ageing mode without FBM; i.e., $n = \infty$ and ageing quotient $q_a = 1.25$, so that each failure is followed by CM and the subsequent system lifetime has an increasingly worse failure rate, in accordance with Formulas (10)–(12);

2. A traditional ageing mode, where ageing is due to an increasing Weibull failure rate. This ageing process can be denoted as theoretical ageing, where each failure is followed by a standard CM intervention that makes the system as good as new, such that the system is replaced ($n = 1$).

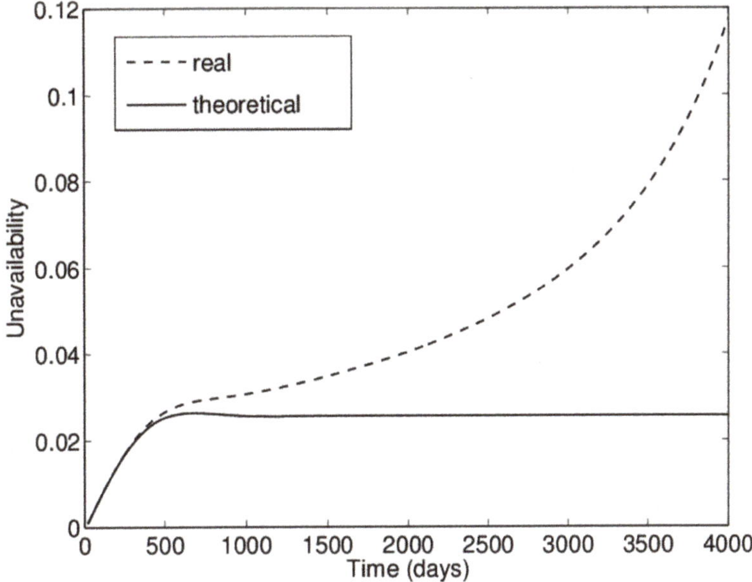

Figure 4. Theoretical versus real ($q_a = 1.25$) ageing of one-component system discussed.

In the first mode, we can see a rapid rise in the unavailability function by the end of mission time, whereas, in the second mode, the unavailability is stabilized shortly after 500 days, which is close to the mean lifetime of the system (531.7 days). These two curves are limiting curves for optimization. The unavailability growth of the real ageing mode can be reduced using a properly selected FBM; i.e., with a properly selected value for the decision variable n respecting a selected unavailability restriction.

Let us take into account that the maximal permissible value of $U_S(x)$ is $U_0 = 0.04$. It is necessary to find an optimal system mode for the objective function given by Equation (1); i.e., we look for an optimal value for the decision variable n. To solve this optimization problem, we must provide data related to the maintenance cost: the cost of one FBM action is $C_R = 12$ and the cost of one CM action is $C_{CM} = 6$, which is in good agreement with practical experience because FBM entails the replacement of the old system with a new one, which is more expensive than repairing it. The results of our optimization process are shown in Table 1, which gives the clear conclusion that the optimal value for the decision variable $n = 5$ and the corresponding minimal cost based on Formula (9) is 59.97 units. Figure 5 shows the dependence of unavailability on time in the optimal mode respecting the given restriction, as well as its comparison with all of the computed system modes.

Table 1. Results for the optimization process for decision variable n in a one-component system.

n	$U_S(x)$	Cost
1	0.026	85.98
2	0.026	64.36
3	0.027	60.82
4	0.029	63.36
5	**0.031**	**59.97**
6	0.034	62.65
7	0.036	65.40
8	0.039	68.22
9	0.042	71.12

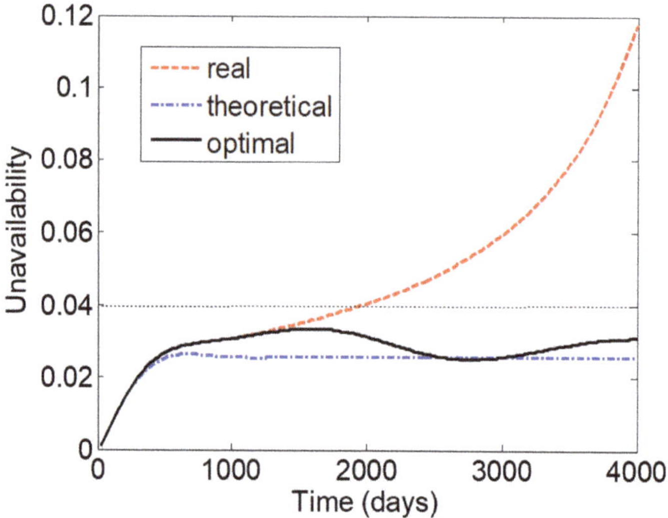

Figure 5. Comparison of courses of the unavailability function $u(t)$ for three different system modes. The optimal value for the decision variable $n = 5$ respecting the given unavailability restriction $U_0 = 0.04$.

The results in Table 1 show that the anticipated gradual growth in unavailability depends on the increasing value of the decision variable n. On the other hand, the cost shows a decreasing trend up to the value of the decision variable $n = 5$ (except for $n = 4$) and an increasing trend for the greater values of n. This outcome can easily be explained:

the increasing value of the decision variable n results in an increasing number of failures per mission time n_R, but the increase in n is faster. For small values of n, it holds true that $n_R > n$, which means that FBM replacements have a direct influence on cost. When n approaches n_R, after a certain value of n is reached, the number of FBM replacements is constant and equal to 1, such that the total cost increases successively only at the expense of the repair cost. If n exceeds $n_R (n > n_R)$, FBM replacement is not achieved; thus, it does not affect the total cost. For example, the decision variable $n = 2$ produces $n_R = 7.7$, such that three FBM replacements influence the total cost, whereas for $n = 5, \ldots, 9$, n_R ranges from 9.0 to 10.85, which means that all of these modes have only one FBM replacement and the total cost only changes due to repairs.

We would like to remark that the cost optimization resulting from Formula (7) can be significantly influenced by other parameters—for example, C_R, C_{CM}—as well as the mission time T_M ([21]), which has a decisive effect on n_R. These parameters are fixed in the analysis and the only decision variable is the number of failures until replacement n.

5.2. Unavailability Quantification and Discrete Maintenance Optimization of a Four-Component System

The second system selected for optimization using the discrete maintenance model based on FBM is a series-parallel system, which consists of two subsystems connected in series. Each subsystem consists of two components connected in parallel, which is demonstrated in Figure 6. The system has been used many times by different authors to demonstrate an imperfect maintenance model; for example, in [7,32], etc. The system parameters and maintenance characteristics are shown in Table 2. The distribution function of the first failure time X_1 of all components is a Weibull distribution, with the shape parameter $\beta = 2$ and scale parameter $\alpha = 1500$ h for both components of the first subsystem, and $\alpha = 2000$ h for both components of the second subsystem. Imperfect CM is characterized by random repair times with a rectangular distribution and a CM MTTR = 300 h for both components of the first subsystem and an MTTR = 200 h for both components of the second subsystem. The replacement time in the FBM model is deterministic, with durations of 75 h for both components of the first subsystem and 50 h for both components of the second subsystem. Similarly to the first example, the replacement time for all components is shorter than the CM time since it can be scheduled beforehand. The imperfect CM model is characterized by a real ageing process, as described above. We presume that the real ageing mode of all components is characterized by the same ageing quotient $q_a = 1.25$, so that each component failure (prior to the nth) is followed by CM and the subsequent component lifetime has an increasingly worse failure rate, in accordance with Formulas (10)–(12). The system is investigated with the mission time $T_M = 8000$ h. Table 2 shows the FBM costs C_R and CM cost C_{CM} of all components.

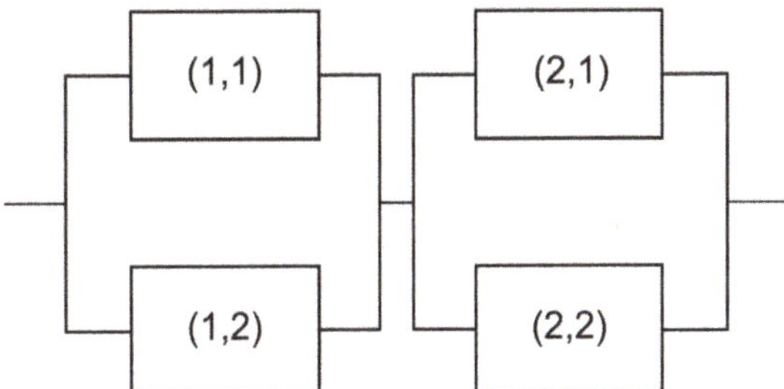

Figure 6. Series-parallel system.

Table 2. System parameters and maintenance characteristics.

Component	α (h)	β	MTTR (h)	Replacement Time (h)	C_R	C_{CM}
(1,1)	1500	2	300	75	12	6
(1,2)	1500	2	300	75	12	5
(2,1)	2000	2	200	50	14	5
(2,2)	2000	2	200	50	15	6

First, the unavailability functions of all components (terminal nodes) were computed according to Formula (6). Then, the system from Figure 6 was transformed into a corresponding AG using the methodology described in [17]. Thereafter, the system unavailability function was calculated using the AG. The difference between real and theoretical ageing is shown in Figure 7, which illustrates the time-dependent unavailability function of the system $u(t)$ with the mission time of 8000 h for two system configurations:

1. All components (terminal nodes) are in a real ageing mode without FBM; i.e., $n = \infty$, each failure is followed by CM, and the subsequent lifetime has a correspondingly increased failure rate.
2. All components are in the traditional ageing mode (theoretical), where ageing is due to an increasing Weibull failure rate and each failure is followed by CM, which brings the system to an as-good-as-new state, such that components are replaced ($n = 1$).

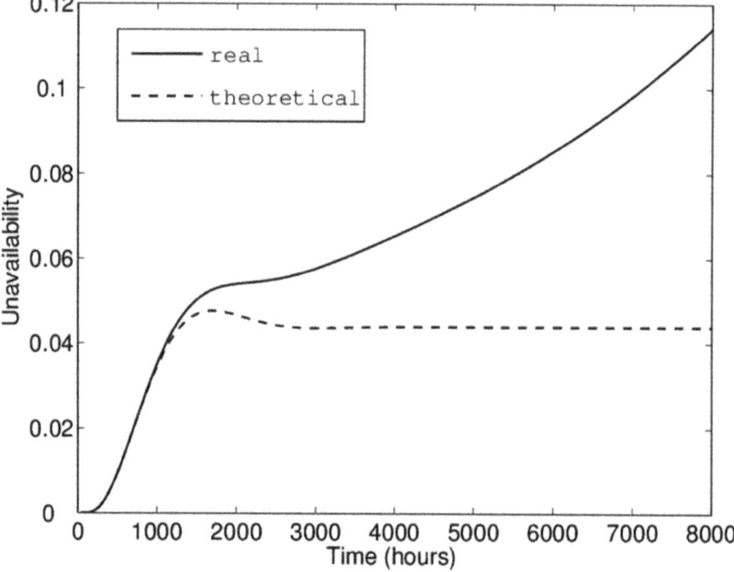

Figure 7. Theoretical versus real ($q_a = 1.25$) ageing of the four-component system.

Not surprisingly, we see a similar effect as was identified for the previous one-component system. In the first system configuration, we could see a rapid rise in the unavailability function, whereas, in the second configuration, the unavailability function peaks at 0.0476 for about 1710 h, which is close to the mean lifetimes of both components of the second subsystem (1772.4 h), and thereafter becomes stabilized. These two curves are, again, limiting curves for optimization. The unavailability growth of the real ageing configuration can be reduced through properly selected FBM for all components. Let us take into account the unavailability restriction that states that the maximal permissible value of $U_S(\mathbf{x})$ is $U_0 = 0.08$. It is necessary to obtain an optimal system configuration for the objective function given by Equation (1) by applying our discrete maintenance model.

We search for optimal component modes; i.e., values of the decision variable n for all four components. In the preliminary calculations, we found that the optimal choice of decision variables ranges from $n = 6$ to $n = 8$. As our system is composed of four components and each component has three maintenance modes, we have to investigate $3^4 = 81$ maintenance configurations of the system. The results of this optimization process are shown in Table 3.

Table 3. Results for optimization process for decision variable n in a four-component system.

	$U_S(x)$	Cost	Index		$U_S(x)$	Cost	Index		$U_S(x)$	Cost	Index
1	0.07168	135.67	1111	28	0.07539	134.776	3321	55	0.08235	126.4	1132
2	0.07189	137.01	2111	29	0.07539	136.016	3312	56	0.08235	126.63	1123
3	0.07189	136.78	1211	30	0.0757	132.34	2131	57	0.08344	127.74	2132
4	0.07191	138.366	3111	31	0.0757	132.11	1231	58	0.08344	127.51	1232
5	0.07191	137.92	1311	32	0.0757	133.81	2113	59	0.08344	127.97	2123
6	0.07211	138.12	2211	33	0.0757	133.58	1213	60	0.08344	127.74	1223
7	0.07214	139.476	3211	34	0.0758	133.696	3131	61	0.08377	129.096	3132
8	0.07214	139.26	2311	35	0.0758	133.25	1331	62	0.08377	128.65	1332
9	0.07217	140.616	3311	36	0.0758	135.166	3113	63	0.08377	129.326	3123
10	0.07441	129.83	1121	37	0.0758	134.72	1313	64	0.08377	128.88	1323
11	0.07441	131.07	1112	38	0.07627	133.45	2231	65	0.08472	128.85	2232
12	0.0748	131.17	2121	39	0.07627	134.92	2213	66	0.08472	129.08	2223
13	0.0748	130.94	1221	40	0.0764	134.806	3231	67	0.08513	130.206	3232
14	0.0748	132.41	2112	41	0.0764	134.59	2331	68	0.08513	129.99	2332
15	0.0748	132.18	1212	42	0.0764	136.276	3213	69	0.08513	130.436	3223
16	0.07487	132.526	3121	43	0.0764	136.06	2313	70	0.08513	130.22	2323
17	0.07487	132.08	1321	44	0.07653	135.946	3331	71	0.08558	131.346	3332
18	0.07487	133.766	3112	45	0.07653	137.416	3313	72	0.08558	131.576	3323
19	0.07487	133.32	1312	**46**	**0.07975**	**125.23**	1122	73	0.08716	127.8	1133
20	0.07520	131	1131	47	0.08053	126.57	2122	74	0.08876	129.14	2133
21	0.07520	132.47	1113	48	0.08053	126.34	1222	75	0.08876	128.91	1233
22	0.07523	132.28	2221	49	0.08072	127.926	3122	76	0.08937	130.496	3133
23	0.07523	133.52	2212	50	0.08072	127.48	1322	77	0.08937	130.05	1333
24	0.07531	133.636	3221	51	0.08141	127.68	2222	78	0.09064	130.25	2233
25	0.07531	133.42	2321	52	0.08164	129.036	3222	79	0.0914	131.606	3233
26	0.07531	134.876	3212	53	0.08164	128.82	2322	80	0.0914	131.39	2333
27	0.07531	134.66	2312	54	0.08189	130.176	3322	81	0.09225	132.746	3333

The sequence of the index digits in Table 3 corresponds to the numbering of the components (2,2), (2,1), (1,2), and (1,1) and determines the component maintenance modes; i.e., values for the decision variable n that range from 6 to 8 (the corresponding digits range from 1 to 3). For example, the sequence index at row number 4 describes such a maintenance configuration, where component (2,2) has the decision variable $n = 8$ (denoted by the number 3 in the sequence), whereas all the components (2,1), (1,2), and (1,1) have a maintenance mode with the decision variable $n = 6$ (all denoted by the number 1 in the sequence).

The values for maximal system unavailability $U_S(x)$ in Table 3 are ordered in ascending order so that it can be seen that the last value of $U_S(x)$, which is below the restriction $U_0 = 0.08$, is in row number 46 and the minimal cost of the first 46 calculations computed according to Formula (9) is 125.23, which is also found in row number 46. This represents the optimal system configuration with the following optimal values for the decision variable n: $n = 6$ for components (2,1) and (2,2) and $n = 7$ for components (1,1) and (1,2). A comparison of the unavailability function of the optimal FBM and the unavailability limiting curves is shown in Figure 8.

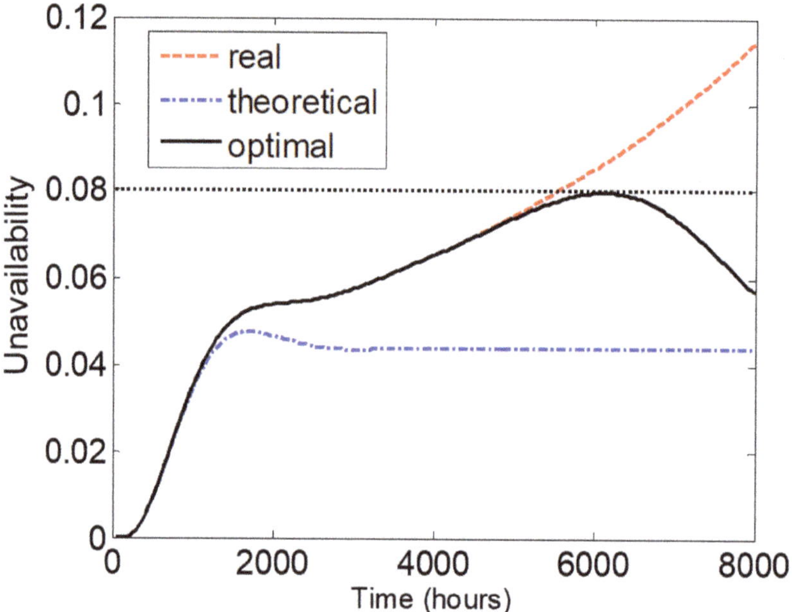

Figure 8. The unavailability function of the optimal FBM in a four-component system, $U_0 = 0.08$.

FBM replacements affect both unavailability and cost. We can further see in Table 3 that the cost of all configurations ranges from 125.2 to 140.6 cost units. Low unavailability values in the first section of Table 3 are caused by frequent FBM replacements—see the often-repeated index 1, which means a minimal value for the decision variable $n = 6$. This effect is particularly noticeable in the last two columns of the index sequence, which correspond to components (1,1) and (1,2). Frequent FBM replacements produce excessive total costs that vary between 130–140 units. On the other hand, in the last section of the Table 3 indexes, 2 and 3 are most common (see again the indexes corresponding to components (1,1) and (1,2)), which implies system configurations with higher maintenance modes; i.e., component modes with higher decision variables $n = 7$ and 8. These values for the decision variable for components (1,1) and (1,2) exceed the mean number of failures per mission time n_R such that FBM replacement is not achieved in most cases, and a relevant part of the total cost of these configurations is at the expense of repair costs, which are lower than replacement costs (Table 2). As a consequence of this, the decreasing total cost mostly varies between 126–131 units, whereas the maximal system unavailability $U_S(\mathbf{x})$ increases to a value of 0.0923.

One more observation can be added. It was mentioned above that components (1,1) and (1,2) have a higher impact on the decision-making process related to the optimization problem. One more reason for this is that the mean time to failure of these two components is less than the mean time to failure of the remaining two components (compare 1329 versus 1772 h). As the first block in the series-parallel system in Figure 6 has greater unavailability than the second block, it is natural that it is of greater importance to solve the optimization problem related to all systems.

6. Conclusions

In this paper, we introduced Weibull-based ageing systems that undergo discrete maintenance optimization. The Weibull-based ageing process is considered imperfect CM, where each failure and follow-up repair degrade the system to some extent. After the occurrence of the n-th failure, where n can be determined for each component as an optimal value for the decision variable of the optimization process, FBM is launched and the component is replaced with a new one. The optimization is realized in a context with minimal system costs and a prescribed unavailability restriction. The corresponding reliability mathematics for the unavailability quantification of terminal nodes was derived in this article because this unavailability is used as the main input in system unavailability quantification using AGs as the system representation. Although the renewal process model that originates from the new renewal cycle was only developed in this article, in general, we can conclude that it is a limited case of the model developed in [31] where the length of the inspection period approaches zero. A cost model respecting the imperfect CM process with FBM was further introduced because it is indispensable for discrete maintenance optimization.

Numerical experiments showed that the discrete maintenance optimization method is a viable method to make an optimal decision for FBM—replacement of system components that undergo real ageing. Although the computing process may in some cases be heavy on computing time (see the 81 system configurations of the four component systems), none of the computations in this paper exceeded 4 min. All these computing experiments, including both the development of the algorithm for the unavailability function $u(t)$ and the discrete maintenance optimization, were numerically realized with the high-performance language MATLAB on computing equipment with the following characteristics: Intel (R) Core™ i7-3770 CPU @ 3.4 GHz and 3.9 GHz, 8.00 GB RAM.

Our future research work will be based on our original achievements resulting from our cooperation with power industry experts ([17]). It will involve maintenance optimization of power networks with a particular focus on circuit breakers.

Author Contributions: Conceptualization, R.B. and P.J.; methodology, R.B. and P.J.; software and validation, R.B.; formal analysis and investigation, R.B. and P.J.; writing—original draft preparation, R.B. and P.J.; supervision, R.B.; funding acquisition, R.B. All authors have read and agreed to the published version of the manuscript.

Funding: This work was partly supported from ERDF „A Research Platform Focused on Industry 4.0 and Robotics in Ostrava Agglomeration", No. CZ.02.1.01/0.0/0.0/17_049/0008425 and partly by the VSB-Technical University of Ostrava project "Applied Statistics and Probability", No.SP2020/46.

Informed Consent Statement: Not applicable.

Data Availability Statement: Not applicable.

Acknowledgments: This work was partly supported from ERDF „A Research Platform Focused on Industry 4.0 and Robotics in Ostrava Agglomeration", No. CZ.02.1.01/0.0/0.0/17_049/0008425 and partly by the VSB-Technical University of Ostrava project "Applied Statistics and Probability", No.SP2020/46.

Conflicts of Interest: The authors declare no conflict of interest.

Glossary

CM	corrective maintenance
PM	preventive maintenance
FBM	failure-based preventive maintenance
AG	directed acyclic graph
C_S	total maintenance cost of a system configuration

$f(\mathbf{x}) = \min C_S$	objective function
$U(\mathbf{x},t)$	instantaneous time-dependent unavailability function
$U_S(\mathbf{x})$	maximal system unavailability within a mission time T_M
U_0	a specified limitation of U_S (maximal permissible value)
$\mathbf{x} = (x_1, \ldots, x_k) \in \mathbf{R}^k$	decision variable
n_i	number of failures of the i-th component, which starts its renewal
k	number of system components
X_1	the time from the beginning of the renewal cycle to the first failure
X_i	the time from the occurrence of the i-th failure to the end of its CM repair (for $i \in \{1, \ldots, n-1\}$)
Y_i	the time from the occurrence of the i-th failure to the end of its CM repair (for $i \in \{1, \ldots, n-1\}$)
Y_n	the time from the occurrence of the n-th failure to the end of the FBM replacement
$X_i^* = X_1 + Y_1 + \cdots + X_{(i-1)} + Y_{(i-1)} + X_i$	the time from the beginning of the renewal cycle to the occurrence of the i-th failure, $X_1^* = X_1$, $X_2^* = X_1 + Y_1 + X_2$, etc.
$F_X(t) = P(X \leq t)$	the cumulative distribution function of the random variable X
$\overline{F}_X(t) = 1 - F_X(t)$	the reliability function of the random variable X

References

1. Ding, S.H.; Kamaruddin, S. Maintenance policy optimization—Literature review and directions. *Int. J. Adv. Manuf. Technol.* **2015**, 5, 1263–1283. [CrossRef]
2. Lee, J.; Lapira, E.; Bagheri, B.; Kao, H.A. Recent advances and trends in predictive manufacturing systems in big data environment. *Manuf. Lett.* **2013**, 1, 38–41. [CrossRef]
3. Pham, H.; Wang, H. Imperfect maintenance. *Eur. J. Oper. Res.* **1996**, 94, 425–438. [CrossRef]
4. Finkelstein, M.; Ludick, Z. On some steady-state characteristics of systems with gradual repair. *Reliab. Eng. Syst. Saf.* **2014**, 128, 17–23. [CrossRef]
5. Liu, Y.; Huang, H.Z. Optimal selective maintenance strategy for multi-state systems under imperfect maintenance. *IEEE Trans. Reliab.* **2010**, 59, 356–367. [CrossRef]
6. Nakagawa, T. Sequential imperfect preventive maintenance policies. *IEEE Trans. Reliab.* **1988**, 37, 295–298. [CrossRef]
7. Pandey, M.; Zuo, M.J.; Moghaddass, R.; Tiwari, M.K. Selective maintenance for binary systems under imperfect repair. *Reliab. Eng. Syst. Saf.* **2013**, 113, 42–51. [CrossRef]
8. Lin, D.; Zuo, M.J.; Yam, R.C. Sequential imperfect preventive maintenance models with two categories of failure modes. *Nav. Res. Logist.* **2001**, 48, 172–183. [CrossRef]
9. Shafiee, M.; Chukova, S. Maintenance models in warranty: A literature review. *Eur. J. Oper. Res.* **2013**, 229, 561–572. [CrossRef]
10. Zhang, M.; Gaudoin, O.; Xie, M. Degradation-based maintenance decision using stochastic filtering for systems under imperfect maintenance. *Eur. J. Oper. Res.* **2015**, 245, 531–541. [CrossRef]
11. Castro, I. Imperfect maintenance: A review. In *Maintenance Modelling and Applications*; Andrews, J., Bérenguer, C., Jackson, L., Eds.; Det Norske Veritas: Bærum, Norway, 2011; pp. 237–262.
12. Morimura, H. On some preventive maintenance policies for ifr. *J. Oper. Res. Soc. Jpn.* **1970**, 12, 94–124.
13. Pulcini, G. *Handbook of Reliability Engineering: Mechanical Reliability and Maintenance Models*; Springer: London, UK, 2003; pp. 317–348.
14. Nakagawa, T. *Maintenance Theory of Reliability*; Springer Science & Business Media: Berlin, Germany, 2006.
15. Drholec, J.; Goňo, R. Reliability database of industrial local distribution system. *Adv. Intell. Syst. Comput.* **2016**, 451, 481–489. [CrossRef]
16. Pinciroli, L.; Baraldi, P.; Zio, E. Maintenance optimization in Industry 4.0. Manuscript Draft JRESS-D-22-01170. *Reliab. Eng. Syst. Saf. in print.*
17. Briš, R.; Byczanski, P.; Goňo, R.; Rusek, S. Discrete maintenance optimization of complex multi-component systems. *Reliab. Eng. Syst. Saf.* **2017**, 168, 80–89. [CrossRef]
18. Badia, F.G.; Berrade, M.D. Optimal inspection of a system under unrevealed minor failures and revealed catastrophic failures. In *Safety and Reliability for Managing Risk*; Soares, C.G., Zio, E., Eds.; Taylor & Francis Group: London, UK, 2006; Volume 1, pp. 467–474, ISBN 0-415-41620-5.
19. Badia, F.G.; Berrade, M.D. Optimum maintenance of a system under two types of failure. *Int. J. Mater. Struct. Reliab.* **2006**, 4, 27–37.
20. Badia, F.G.; Berrade, M.D. Optimum maintenance policy of a periodically inspected system under imperfect repair. *Adv. Oper. Res.* **2009**, 13, 691203. [CrossRef]

21. Badia, F.G.; Berradea, M.D.; Lee, H. An study of cost effective maintenance policies: Age replacement versus replacement after N minimal repairs. *Reliab. Eng. Syst. Saf.* **2020**, *201*, 106949. [CrossRef]
22. Sheu, S.H.; Tsai, H.N.; Wang, F.K.; Zhang, Z.G. An extended optimal replacement model for a deteriorating system with inspections. *Reliab. Eng. Syst. Saf.* **2015**, *139*, 33–49. [CrossRef]
23. Briš, R. Parallel simulation algorithm for maintenance optimization based on directed Acyclic Graph. *Reliab. Eng. Syst. Saf.* **2008**, *93*, 852–862. [CrossRef]
24. Briš, R. Exact reliability quantification of highly reliable systems with maintenance. *Reliab. Eng. Syst. Saf.* **2010**, *95*, 1286–1289. [CrossRef]
25. Weide, J.A.M.; Pandey, M.D. A stochastic alternating renewal process model for unavailability analysis of standby safety equipment. *Reliab. Eng. Syst. Saf.* **2015**, *139*, 97–104. [CrossRef]
26. Vaurio, J. Unavailability of components with inspection and repair. *Nucl. Eng. Des.* **1979**, *54*, 309–324. [CrossRef]
27. Vaurio, J. On time-dependent availability and maintenance optimization of standby units under various maintenance policies. *Reliab. Eng. Syst. Saf.* **1997**, *56*, 79–89. [CrossRef]
28. Vaurio, J. Availability and cost functions for periodically inspected preventively maintained units. *Reliab. Eng. Syst. Saf.* **1999**, *63*, 133–140. [CrossRef]
29. Caldarola, L. Unavailability and failure intensity of components. *Nucl. Eng. Des.* **1977**, *44*, 147–162. [CrossRef]
30. Cui, L.; Xie, M. Availability of a periodically inspected system with random repair or replacement times. *J. Stat. Plan. Inference* **2005**, *131*, 89–100. [CrossRef]
31. Jahoda, P.; Bris, R. A renewal process model with failure based PM and imperfect CM for unavailability exploration. *Int. J. Qual. Reliab. Manag.* **2021**, *39*, 984–999, *Advance online publication.* [CrossRef]
32. Cassady, C.R.; Pohl, E.A.; Murdock, W.P. Selective maintenance modeling for industrial systems. *J. Qual. Maint. Eng.* **2001**, *7*, 104–117. [CrossRef]

Article

Deep Machine Learning Model-Based Cyber-Attacks Detection in Smart Power Systems

Abdulaziz Almalaq [1,*], Saleh Albadran [1] and Mohamed A. Mohamed [2,*]

1   Department of Electrical Engineering, Engineering College, University of Ha'il, Ha'il 55476, Saudi Arabia; s.abadran@uoh.edu.sa
2   Electrical Engineering Department, Faculty of Engineering, Minia University, Minia 61519, Egypt
*   Correspondence: a.almalaq@uoh.edu.sa (A.A.); dr.mohamed.abdelaziz@mu.edu.eg (M.A.M.)

Abstract: Modern intelligent energy grids enable energy supply and consumption to be efficiently managed while simultaneously avoiding a variety of security risks. System disturbances can be caused by both naturally occurring and human-made events. Operators should be aware of the different kinds and causes of disturbances in the energy systems to make informed decisions and respond accordingly. This study addresses this problem by proposing an attack detection model on the basis of deep learning for energy systems, which could be trained utilizing data and logs gathered through phasor measurement units (PMUs). Property or specification making is used to create features, and data are sent to various machine learning methods, of which random forest has been selected as the basic classifier of AdaBoost. Open-source simulated energy system data are used to test the model containing 37 energy system event case studies. In the end, the suggested model has been compared with other layouts according to various assessment metrics. The simulation outcomes showed that this model achieves a detection rate of 93.6% and an accuracy rate of 93.91%, which is greater compared to the existing methods.

Keywords: cyber-attack detection; deep machine learning; smart power grid; data processing

MSC: 94-10

Citation: Almalaq, A.; Albadran, S.; Mohamed, M.A. Deep Machine Learning Model-Based Cyber-Attacks Detection in Smart Power Systems. *Mathematics* **2022**, *10*, 2574. https://doi.org/10.3390/math10152574

Academic Editors: Gurami Tsitsiashvili and Alexander Bochkov

Received: 6 June 2022
Accepted: 22 July 2022
Published: 25 July 2022

Publisher's Note: MDPI stays neutral with regard to jurisdictional claims in published maps and institutional affiliations.

1. Introduction

1.1. Necessity of the Research

Cyber-physical systems (CPS) attempt to couple the physical and cyber-worlds, and they are extensively employed by industrial control systems (ICS) to provide users with all the data they need in real-time [1]. Power distribution systems and waste-water treatment plants are among the areas where CPS is being used. Nevertheless, CPS security problems differ from conventional cyber-security problems in that they include integrity, confidentiality, and availability. In addition to transmitting, distributing, monitoring, and controlling electricity, a smart grid (SG) would greatly enhance energy effectiveness and reliability. Such systems may fail and result in temporary damage to infrastructures [2]. Power grids are regarded as essential infrastructure nowadays by many societies, which have developed security measures and policies related to them [3]. Phasor measurement units (PMUs) are adopted in modern electrical systems to improve reliability as they become more complex in their structure and design. Utilizing the gathered information for quick decision making is one of the advantages. There is still the possibility that hacker exploits vulnerabilities to result in branch overloaded tripping, which will lead to cascading failures and, therefore, leads to considerable damage to SG systems [4]. As the operators monitor and manage the energy grid, they must consider possible attacks on the grid. To accomplish this, much energy and grid expertise is required. However, deep machine learning (DML) methods are used because of their capability to recognize patterns and learn, as well as being quickly able to identify potential security boundaries [5].

1.2. Literature Review

Network systems, usually referred to as essential infrastructure systems, have been usually applied to link the systems for monitoring and collecting equipment operations in real-time. The supervisory control and data acquisition (SCADA) system is highly vulnerable to cyber-attacks, and such attacks need to be handled with extreme caution [6]. Sensor's fingerprints and noise processing are used in [7] for detecting hidden cyber-attacks in CPS, and the data set from the actual-world water treatment plants is employed to validate the approach, and the outcomes indicated an accuracy of 98%. In [8], a semantic instruction detection system on the basis of the network was examined for detecting attacks on water plant processes by analyzing network traffic. These findings highlight the need for CPS investigation. Cyber and physical systems are part of the SG. Intrusion detection problems are solved using DML, as seen in recent research [9–11]. The intrusion detection method on the basis of DML is examined in [9]. The data set employed was a SWAT-produced datum from various attacks of 10 various kinds. A quick one-class classification scheme that overcomes the problem of vast sensitivity to out-of-range data is employed in [10], and an actual data set is used to test the suggested algorithm. The data sets employed in this study have also been utilized in numerous other types of research. The authors in [11] examined the method with accuracy rates of around 90% for JRipper + Adaboost and 75% for random forest compared to the whole multiclass data set. The privacy preservation intrusion diagnosing method on the basis of the correlation coefficient and expectation maximization (EM) clustering techniques is presented in [12] to select significant sections of data and recognize intrusive occurrences. There was an 88.9% recall rate in the model compared to the multiclass data sets with 75% of features. Authors in [13] have improved the detection process by dropping the defense target from rejecting attacks to preventing outages to decreasing the necessary number of secured PMUs. In [14], the authors investigated the effect of cyber-attack on the PMU state estimation process using the Cartesian equations and in the case of zero injection buses. In [15], it is tried to develop an allocation method for fault observability using PMU data considering zero injection buses. In [16], the authors have introduced a fault detecting and classifying, and placement approach based on advanced machine learning in radial distribution systems.

1.3. Contributions

A model based on machine learning is presented in this study for detecting system behaviors by analyzing historical data and related log data. Although unsupervised learning is beneficial for detecting zero-day attacks since it requires no training in attack scenarios, it is also vulnerable to false positives [17]. Furthermore, supervised learning can clearly improve the detection's confidence. The experiments are then performed using the supervised machine learning approach. The main contributions in this paper are summarized as follows:

(1) Feature construction engineering is performed, and 16 novel features are constructed via an analysis of the features and possible links of the raw data in the electrical network. It is possible to construct novel features using a combination of attributes that could help more effectively utilize possible types of data instances, which could be used in machine learning models for better application.

(2) A new process for handling abnormal data, such as not the number and infinity amounts in the data sets, is proposed. The suggested approach could significantly enhance accuracy in comparison to conventional processes of processing abnormal data.

(3) A classification model based on machine learning is constructed. The average accuracy of 0.9389, precision of 0.938, recall of 0.936, and F1 score of 0.935 on 15 data sets demonstrate that the suggested model successfully distinguished 37 kinds of behaviors such as power grid fault and single-line-to-ground (SLG) fault replay, relay setting varies, and trip command injection attacks.

Following are the remaining sections of the study. A detailed explanation of the methodology is provided in Section 2. The results of the classification are discussed in Section 3. The conclusion appears in Section 4.

2. Model Structure

Scenarios where disturbances and attacks happen in the electric grid, as well as the meaning of features in the data set, are presented in this part. The suggested model and data processing are detailed here.

2.1. Introduction to Power System Framework Configuration

The suggested data set consisting of measurements associated with normal, fault, and cyber-attack behavior, and so on [18–20]. The electrical network block diagram is shown in Figure 1 [21]. Relay, control panel, snort, and PMU/synchronous are primarily used for recording measurement data. Following are some of the most significant components. Power generators are shown by P1 and P2, and the intelligent electronic device (IED) is relay R1, which could switch breaker1 (BR1) on or off. Transmission lines (TLs) are represented by L1 and L2. The phasor data concentrator is shown by PDC that stores and displays Synchron-phasor data as well as records historical data. The IED incorporates a distance protection mechanism that can trip the breaker if it detects faults. Due to the absence of internal verification approaches for detecting changes, the breaker will be tripped regardless of whether the fault is valid or not. BR1-4 can be tripped by manually sending relevant commands to IEDs. In the event that lines or other components are to be maintained, the manual override will be necessary.

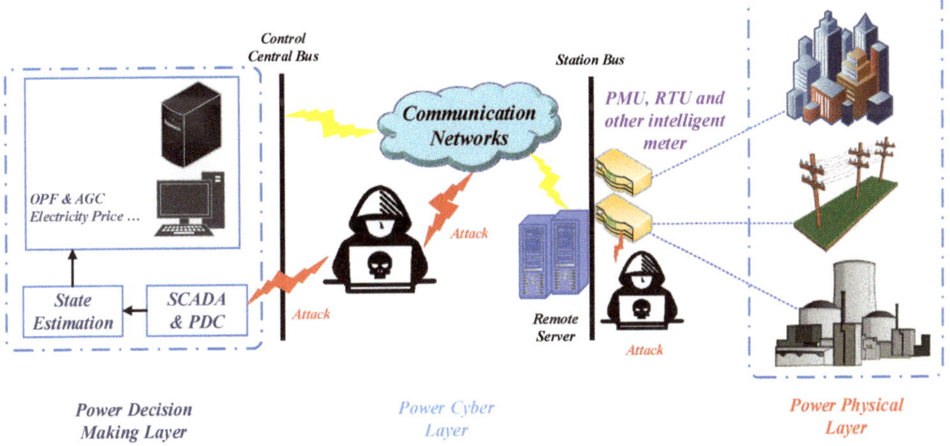

Figure 1. The power system framework configuration.

This experiment applied a data set that contains 128 features recorded using PMUs 1 to 4 and relay snort alarms and logs (Relay and PMU have been combined). A synchronous phasor, or PMU, measures electric waves on a power network using a common time source. A total of 29 features could be measured by every PMU. The data set also contains 12 columns of log data from the control panel and one column of an actual tag. There are three main categories of scenarios in the multiclass classification data set: No Events, Events, Intrusion, and Natural Events. Table 1 summarizes the scenarios, and a brief explanation of each category is provided in the data set.

(a)   SLG fault: A fault occurs whenever the current, voltage frequency of the system changes abnormally, and many faults in electrical systems occur in line-to-ground and line-to-line (LL). The simulated SLG faults are represented as short circuits at diverse points along the TL in the data set.

(b)   Line maintenance: This type of attack is caused when one or more relays have been deactivated on a particular line to maintain.

(c)   Data injection: More research is being conducted into false data injection state estimation in electrical networks. False data injection attacks are one of the main forms of network attacks, which could affect the power system estimation method. Attackers alter phase angles in order to create false sensor signals. The objective of such attacks is to blind the operators and to avoid raising an alarm, which could lead to economic or physical damage to the electrical systems. Attackers synchronize the phasor measurement with the fault's SLG and next send a relay trip command on the affected lines. A data set modeled the conditions by varying variables, such as current, voltage, and sequence components, which caused faults on various levels ([10 to 19]%, [20 to 79]%, [80 to 90]%) of the TLs.

(d)   Remote tripping command injection attack: This occurs when a computer on the communications network uses unexpected relay trip commands to relay at the end of a TL. For achieving attacks, command injection has been applied versus single relays (R [1–4]) or double relays (R3 and R4, R1 and R2).

(e)   Relay adjusting variation attack: The relay is configured with a distance protection layout. Attackers change the setting, so the relay responds badly to authentic faults. In the data sets, faults were caused via deactivating the relay functions at diverse parts of TLs with R1 or R2 or R3 or R4 deactivated and fault.

Table 1. Explanation of scenarios.

Case Study No.	41	1–6	13, 14	7–12	15–20	21–30, 35–40
Explanation	Usual operation load variations	SLG faults	Line maintenance	Data injection	Remote tripping command injection	Relay setting vary
Kind	No events	Natural events			Intrusion events	

2.2. Methodology

Despite the fact that the machine learning approach is capable of detecting disturbances and cyber-attacks on electric grids, it can have these drawbacks. Currently, references just discuss how to diagnose attacks in the electrical grids and seldom examine the data relationship. In contrast, when working with multi-classification problems, many algorithms convert them into multi-two-class situations. Nonetheless, the AdaBoost algorithm is able to handle multi-classification situations directly. It utilizes weak classifiers well for cascading and is capable of using various classification algorithms as weak classifiers. In terms of the error rate of misclassification, the AdaBoost algorithm is highly competitive [22]. With an increase in data amount, the fitting ability is affected both by generalization problems and by the increasing difficulty of computing. Machine learning requires a large amount of calculating to find the best solution. Additionally, the accuracy rates on the model presented in [11,12] are about 90% compared to the multiclass data sets, which provides considerable space for development. As a consequence of these findings, this paper constructs a model that can perform superior feature engineering and next can split the data by the diverse PMUs to minimize computation overhead. It should be noted that the PMU allocation in the smart grid is performed in the planning stage and might be implemented according to different purposes. While the high cost might be a limitation, the high number of PMUs is always preferred to cover all areas of the smart grid. It is worth noting that PMU allocation is out of the scope of this work but can be found in other research works widely. In addition, the AdaBoost algorithm for detecting the 37-class fault and cyber-attack case studies in the electric grids is adopted in this paper.

About the feature selection process, it should be noted that this experiment applied a data set that contains 128 features recorded using PMUs 1 to 4 and relay snort alarms and logs (relay and PMU have been combined). Please also note that each PMU can record 29 different features. In this regard, and in order to obtain enriched and integrated informative data, feature construction engineering is performed, and 16 novel features are constructed via an analysis of the features and possible links of the raw data in the electrical network. Technically, it is possible to construct novel features using a combination of attributes that could help more effectively utilize possible types of data instances, which could be used in machine learning models for better application. It is worth noting that we made use of the random forest method to create and classify features. Finally, based on anticipation weighted voting, 37 various case studies were implemented for simulation purposes.

2.3. Diagnosing Attack Behavior Model Structure

A model architecture diagram is shown in Figure 2 to detect faults and cyber-attack in electrical grids. According to Figure 2, the model architecture usually consists of four stages: property making, data dividing, weight voting, and layout training as follows:

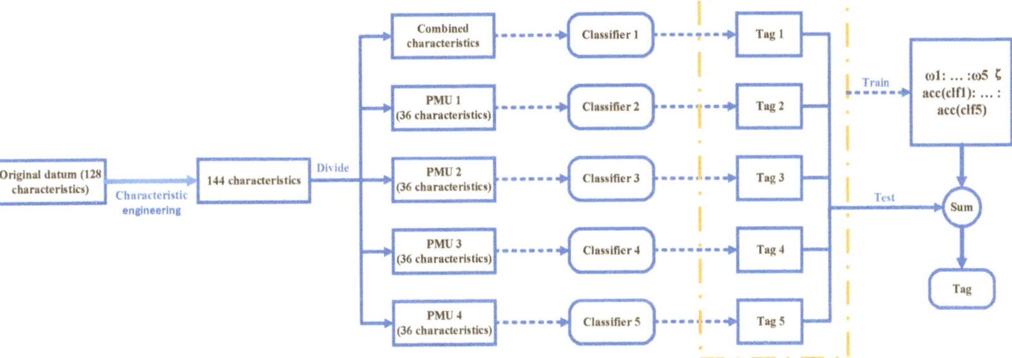

Figure 2. Explanation of layout to detect disturbance and cyber-attack in electrical networks.

Stage.1. Property making. By creating novel features manually from the original data set, it is able to improve the dimension of the data. A novel piece of data is generated by integrating the novel features with several original ones. The upper limit of the model is determined by the features and data, and the algorithm can just approximate the upper limit as closely as feasible. In order to achieve maximum accuracy and improve robustness, feature construction engineering is essential. It is important for feature construction using the original data to obtain more flexible features, and therefore increase data sensitivity and increase the ability to analyze it in the case of sending it to models for classification and training. The target of helpful features is to be simple to understand and maintain. The results of the analysis have led to the construction of 16 novel features. There is also a tendency in machine learning problems to include a large number of features for training instances, and it results in excessive computational overhead and overfitting, leading to poor efficiency. The curse of dimensionality has usually been used to describe this problem. Feature selection and feature extraction have been widely applied to mitigating the problems caused by high dimensionality in learning problems [23].

Stage.2. Datum dividing and training. The test and training sets are divided through 9:1 through the data splitting module. There is too much noise in the classifier if too many features are used [24]; therefore, every original data has been split into four parts according to features from various PMUs. While doing this, a section of the main characteristics is picked and sent to the AdaBoost layout to train alongside the novel features as well. This step is necessary for reducing the effect of errors resulting from bad PMU measurements. In case the feature dimension increases, the classifier's performance decreases. As a result of this step, several of the original features are combined with novel ones in order to reduce the dimension. The original features are sorted using feature importance, and afterward, a variety of proportions of the features are selected, explained in more detail in Part 3. In addition, several classifier models are developed for personalizing the features following splitting. Various classifiers are set up to make every section of the data display the greatest impact on the classifier, i.e., the training model. Using five classifiers and later obtaining five tags following transferring the information to the layout reduces the effect of the alone classifier generalization error.

Stage.3. Weights for voting. It is the responsibility of the module to assign diverse weights to the tags derived from diverse classifiers and vote on the last classification tag of the data. According to the accuracy ratio of every classifier in the training set, the ratio of various weights has been thus determined. Various tags are generated by the test set following they have passed through the trained classifier, and the weights are determined for the last voting session based on the tags of the relevant classifier. By updating the weights in real-time, the entire system can become more robust and generalizable.

2.4. In-Depth Explanation of the Attack-Diagnosing Layout

2.4.1. Properties Making

During property making, 16 novel features have been extracted from every PMU measurement feature and incorporated into the original data set for preparing for the next step. Raw data is mainly used for extracting novel features based on corresponding computations. Table 2 shows the name, explanation, and extraction process of the extracted feature.

Table 2. Explanation of extracted characteristics.

Feature	VCA4	VCA1	SI
Description	PA7:VH-PA 10:IH	Sin (PA1:VH-P4:PA4:IH-PA7:VH-PA10:IH)	Sin (PA4:IH-PA 10:IH)
Feature	**SV**	**VCM1**	**VCM2**
Description	Sin (PA1:VH-PA 7:VH)	(PM1:V-PM7:V)/(PM4:1-PM10:I)	(PM2:V-PM8:V)/(PM5:I-PM11:I)

2.4.2. Data Processing

It is important to process the data prior to sending it to the machine learning model. The normalization of the data is an important part of data processing. The benefit of this method is that it speeds up and improves the accuracy of iterations for finding the best solution for gradient descent. Among the most common techniques of data normalization are z-score standardization and min-max standardization. Basically, min-max standardization works by changing the original data linearly toward an outcome between [0, 1] shown below:

$$X_{scale} = \frac{x - x_{min}}{x_{max} - x_{min}} \tag{1}$$

In addition, Z-score standardization has been known as standard deviation standardization, and it has been mostly applied for characterizing deviations from the average. The data analyzed through this technique assure the standard usual distribution, which is that the standard deviation and average are equal to one and zero, respectively. The data processed using the process can satisfy the standard normal distribution, meaning the mean equals 0 and the standard deviation Equation (1). Following is the transformation function, the mean amount of the instant data is shown by μ, and the standard deviation is represented by σ. This study adopts this normalization process.

$$X_{scale} = \frac{x - \mu}{\sigma} \tag{2}$$

A data set may contain the not a number (NAN) and infinity (INF) amount, but it has been usually substituted through the mean amount or zero. For the data set applied here, the novel replacement process is proposed to avoid underflows in the final replacement value and the data being overly discrete. *log_mean* value is used for replacing NAN and INF values present in the data. It can be calculated as follows:

$$log\_mean = \frac{\sum \log|k_i|}{Num(k_i)} \cdot \left(1 - 2 \triangleleft \left(\frac{\sum k_i}{Num(k_i)} < 0\right)\right) \tag{3}$$

Here, the number of digits in a column is shown by $Num(k_i)$ and the indicator function is represented by $\triangleleft(x)$, which can be described in the following way:

$$\triangleleft(x) = \begin{cases} 1 \ if \ x \ is \ true \\ 0 \ otherwise \end{cases} \tag{4}$$

Comparative experiments are conducted on various treatment approaches in this study. Section 3 shows the outcomes that show that the suggested process succeeds.

2.4.3. Establish Classifier Layouts

During the process of making the classifier scheme, the features and characteristics of the SG information are considered, and various DML classification schemes are established for the data obtained from every PMU. Various experiments have shown that random forest is the best for the data gathered through every PMU, and AdaBoost is the ideal layout for combined features, including a section of the main characteristics as well as properties derived from the property making. With AdaBoost, several basic classifiers are combined into a robust classifier. The experiment proposes a new model in which random forest has been applied as the basic classifier of AdaBoost, followed by weighted voting on the anticipation outcomes (AWV).

Stage. (1) Set the training data's weights of observation $= (\omega_1, \dots \omega_2, \dots \omega_n) \ \omega_i = 1/n$.
Stage. (2) For $t = 1{:}T$

(I) Select random forest classifier $RFC^{(t)}$ as the base classifier of Adaboost.
(II) Calculate classification error $\varepsilon^{(t)} = \sum_{i=1}^{n} \omega_i^{(t)} \triangleleft (y_i \neq RFC^{(t)}(X_i) / \sum_{i=1}^{n} \omega_i^{(t)}$
Here, X_i shows the *i*th input feature vector, the actual tag of the *i*th input property vector is represented by y_i. The predicted outcome is shown through $RFC^{(t)}(X_i)$.
(III) Calculate $\alpha^{(t)} = 0.5 \ln\left(\frac{1-\varepsilon^{(t)}}{\varepsilon^{(t)}}\right)$.
(IV) Update the weights through $\omega_i^{(t+1)} = \omega_i^{(t)} \exp\left(\alpha^{(t)} \triangleleft \left(y_i \neq RFC^{(t)}(X_i)\right)\right)$
(V) Renormalize so that $\sum_{i=1}^{n} \omega_i = 1$.

Stage. (3) Output $C(x) = argmax_y \sum_{t=1}^{T} \alpha^{(t)} \triangleleft (RFC^{(t)}(X) = y$

Here, $argmax_x(f(x))$ function is meaned return the amount of x which maximizes $f(x)$. Here, for 37-class classification problem, so $\in (1, 2, \dots, 37)$, and $\sum_{t=1}^{T} \alpha^{(t)} \triangleleft (RFC^{(t)}(X) = y$ is a 37-dimensional vector. When various probabilities are associated with various tags for one feature vector X_i, the last output is determined through the probability with the highest amount.

2.4.4. Voting with Weights

Hard combination and soft combination are two ways of addressing the final multiple tags [25]. The hard combination is training the similar data set section with various DML methods and assigning the similar weight to the achieved last tags for voting. The result is the tag with the highest weight value. Similar to that, the soft combination involves adopting various DML methods for a similar section of the data set. However, the tags are assigned with different weights, and the end result is the tag with the highest weight. To summarize, the main difference between the hard and soft combinations is whether or not the weights are equal. In a classifier, weights represent the probability value of a tag or its confidence level. The present study sets up various machine learning models for various data blocks to address multi-tag problems so as to make the model perform effectively for the data set. Lastly, different weights are assigned to tags to determine the final results. Algorithm 1 describes these steps.

Algorithm 1: Weight Voting Scheme

Input:	144 characteristics
Output:	Tag
(1)	Divide data by random Num (training set):Num (test set) = 9:1
(2)	Divide 144 characteristics into 4 section $PMU_i\_charectristics$ (i = 1, 2, 3, 4)
(3)	Transfer training set to the various machine learning; layout and take the precision rate acc (clf_i) (i = 1, 2, 3, 4, 5)
(4)	Transfer trail information to the trained layout and produce five tags; $label_i$ (i = 1, 2, 3, 4, 5)
(5)	Initialize weight ω_i (i = 1, 2, 3, 4, 5) and $\omega_1 : \ldots : \omega_5 \approx acc$ (AdaBoost) : acc(RFC_1) : $\ldots$: acc(RFC_4)
(6)	Merge tags with weights $[[abel_1, w_1], \ldots, [abel_5, w_5]]$
(7)	Constitute a tag set (tag), and compute the weight set W regarding the tag in the set
(8)	Chose the tag with the largest weight in the W as the last outcome

---

3. Experiment and Evaluation

In machine learning, classifications and regressions are the primary learning tasks. It is obvious that the classification problem is addressed in this study. The next experiments are designed to test whether the model structure described in this study is capable of distinguishing fault and disturbance in electrical systems. A comparison is made between the model and various conventional models, such as convolution neural network (CNN), gradient boosting decision tree (GBDT), extreme gradient boosting (XGBoost), decision tree (DT), support vector machine (SVM), and k-nearest neighbor (KNN).

Additionally, the accuracy achieved through transferring information has been compared after the property making is compared.

3.1. Data Set

A multiclass classification data set for ICS cyber-attacks is used in the present study. There are a total of 15 groups in the multiclass data set, each with about 5000 pieces of data. Each group's situation is shown in Table 3. Across all tag kinds, the distribution of data can be fairly uniform. ARFF (Attribute-Relation File Format) is the main file template of the data set. An ARFF file is the ASCII text format, which represents a set of attributes shared by several samples. To ease the process, ARFF files are converted to CSV (Comma Separated Values) template. In CSV files, textual/numeric tabular information is stored in plain text. AUC, F1 score, ROC curve, ROC curve, precision, accuracy, and recall area are primarily used to evaluate classification models in machine learning. There are several terms applied in machine learning that require an explanation. The true positive (TP) is the positive sample that the layout predicts to be positive, the false positive (FP) is the negative sample that the layout predicts to be positive, and the false negative (FN) is the positive sample that the model predicts to be negative, the true negative (TN) is the negative sample that the model predicts to be negative. The suggested layout is evaluated using accuracy, precision, recall, and F1 score. An F1 score is basically the harmonic value of precision and recall, which are calculated according to the following equations:

$$accuracy = (TP + TN)/(TP + FP + FN + TN) \tag{5}$$

$$precision = TP/(TP + FP) \tag{6}$$

$$recall = TP/(TP + FN) \tag{7}$$

$$F1\ score = \frac{2TP}{2TP + FN + FP} = \frac{2 \cdot precision \cdot recall}{precision + recall} \tag{8}$$

Table 3. Multiclass instance data statistics.

Data set	Data 1	Data 2	Data 3	Data 4	Data 5	Data 6	Data 7	Data 8
Data number	4966	5069	5415	5202	5161	4967	5236	5315
Data set	Data 9	Data 10	Data 11	Data 12	Data 13	Data 14	Data 15	Entire
Data number	5340	5569	5251	5224	5271	5115	5276	78,377

3.2. Experiment Outcome

3.2.1. Machine Learning Model

In this experiment, KNN, SVM, GBDT, XGBoost, CNN, and others were applied as conventional models.

(A) Based on the distance among feature values, the K-nearest neighbor algorithm has been categorized. Distance is calculated primarily using Euclidean/Manhattan distances formulation.

(B) The SVM [26] layout uses the sample as a spot in the region and applies various mapping functions for mapping the input into the great-dimensional property region for constructing the hyperplane group or hyperplane. According to intuition, the further away the boundary is from the point of data training, the more accurate the classification will be. $\omega^T x + b = 0$ shows the formulation to divide the hyperplane, in which the normal vector is shown by ω determining the hyperplane's direction., and the displacement term is shown by b determining the distance between the hyperplane and the origin. $\gamma = (\omega^T x + b)/||\omega||$ show the formulation for the interval from each spot x to the hyperplane in the region, γ must be maximized within the conditions, which the hyperplane properly divides the training instances, i.e.:

$$\max_{\omega,b} \frac{2}{||\omega||}$$
$$subject\ to\ y_i\left(\omega^T x + b\right) \geq 1 \tag{9}$$

Calculating the limitation problem via the Lagrange function is more efficient, and an objective function can be derived from the following formula, in which α_i shows the Lagrange multiplier and $\alpha_i \geq 0$.

$$L(\omega, b, \alpha) = \frac{1}{2}||\omega||^2 + \sum_{i=1}^{m} \alpha_i\left(1 - y_i\left(\omega^T x + b\right)\right) \tag{10}$$

Determine $L(\omega, b, \alpha)'s$ partial derivatives and make them 0:

$$\frac{\partial L(\omega, b, \alpha)}{\partial \omega} = 0,\ \frac{\partial L(\omega, b, \alpha)}{\partial b} = 0 \tag{11}$$

The dual problem can be as follows:

$$\max_{\alpha} \sum_{i=1}^{m} \alpha_i - \frac{1}{2}\sum_{i=1}^{m}\sum_{j=1}^{m} \alpha_i \alpha_j y_i y_j x_i^T x_i\ subject\ to\ \sum_{i=1}^{m} \alpha_i y_i = 0,\ \alpha_i \geq 0 \tag{12}$$

(C) The decision tree algorithm starts with a group of instances/cases and then makes a tree information framework, which is applied to novel cases. A group of amounts/symbolic amounts describes every case [27]. Entropy is used in C4.5 and C5.0 for the spanning tree algorithm.

(D) A boosting algorithm has been used to improve the XGBoost [28] classifier algorithm. The model is based on residual lifting. Based on the error function, the objective function is calculated by taking the prime and second derivatives of every data spot. The loss function is a square loss. Here is its objective function, in which l shows a differential convertible loss function, which shows variation among the prediction $\hat{y}_i$ and the purpose y_i. The second part Ω can penalize the pattern complexity, and T shows the leaves number in the tree. The γ and λ show the tree's complexity, the greater their amount, and the simpler the framework of the tree.

$$L(\phi) = \sum_i l(\hat{y}_i, y_i) + \sum_k \Omega(f_k) \ \ where \ \Omega(f) = \gamma T + \frac{1}{2}\lambda||\omega||^2 \tag{13}$$

(E) The random forest exhibits excellent efficiency and has been extensively applied [29]. RF utilizes the decision tree as its base classifier and shows an extension of Bagging. RF uses two very significant procedures. The first technique involves introducing random features in the procedure of decision tree making, and the second involves an out-of-bag estimation. The RF method can be described below. The first step is to randomly select a sample from every data, and afterward, to return the sample to the original data. As a root sample for a decision tree, the chosen samples have been applied for training the decision tree. Second, for splitting the nodes of the decision tree, m attributes have been chosen randomly (there are a total of M attributes and ensuring $<< M$). Choose an attribute to be the dividing feature of the node using the strategy, such as information gain. Continue to do this until the decision tree can no longer be divided.

(F) Among the more popular deep learning networks is CNN. There are usually input, output, latent, and max-pooling layers in a CNN model. Several great results have been obtained in numerous areas of computer vision. Here, one-dimension property vectors are used as input, and a one-dimension convolution kernel in convolution layers is adopted. The convolution layer extracts properties from the input, and here the kernel size is three. The process of the CNN model is shown in Figure 3.

Figure 3. The procedure of CNN layout.

Actually, the main purpose of this research is to show the high and successful role of the deep learning models in reinforcing the smart grid against various cyber-attacks. In this regard, the proposed model would detect and stop cyber-hacking at the installation location rather than focusing on the cyber-attack type. Therefore, the localization procedure would be attained through the diverse detection models located in the smart grid, but the cyber-attack type detection requires more data that can be made later based on the recorded abnormal data.

3.2.2. Outcomes

This study considers 37 varied scenarios for events. In order to determine the need for various models (fault analysis), we performed some comparative experiments according to various PMU kinds. In one group, properties of localization/segmentation are sent to the related DML model in order to train, and in the other one, whole features are sent to various machine learning models. Moreover, it is shown in Table 4 that data can be effectively split according to the PMU resources. Splitting the data can enhance the accuracy of classification models as well as reduce data dimensions and enhance training speed and minimize computing sources. The score of the significant features is shown in Figure 4.

Table 4. Transfer diverse characteristics to the layout for comparison.

Technique	Characteristics	
	Entire	**Split**
Accuracy	0.9344	0.9387

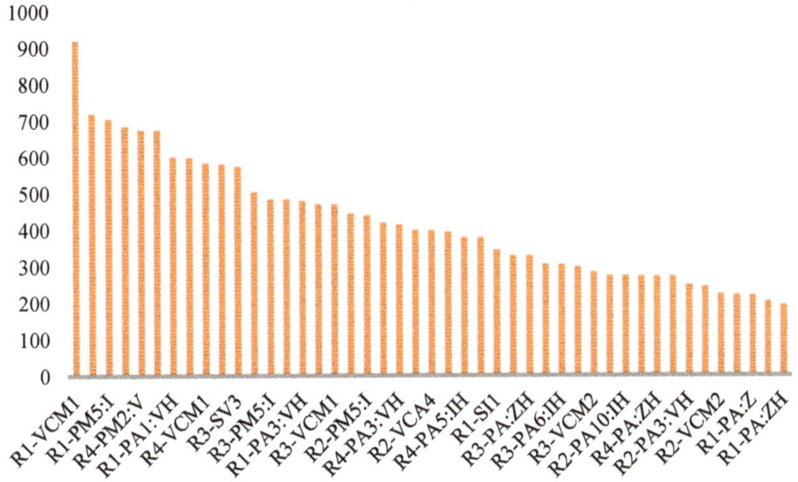

Figure 4. Significance features score.

Several corresponding experiments are conducted on various ways of replacing abnormal values in data. Table 5 shows the outcomes. The replacement method is shown in the left column, and the suggested approach is represented by *log_mean*. Zero shows a process to replace NAN and INF with zero values, and mean shows a process to replace with the mean value. The AWV model is utilized as a trial model, and the accuracy is adopted as the assessment metrics, that is, the right column in Table 5.

Table 5. Diverse methods to procedure Inf and Nan.

Method	Zero	Mean	*Log-Mean*
Accuracy	0.9361	0.9342	0.9387

Applying the *log_mean* technique for replacing the unusual amount in the data is intuitively the best approach. According to the outcome, the suggested process in order to process abnormal values has proven successful.

Comparison experiments are also conducted to verify feature selection. First, the significance of the original features is determined, and afterward, they are arranged based on significance. A variety of mixtures of features has been selected for training, and Table 6 shows these outcomes.

The approach was verified practically through a comparative test. The test extracts the test group and training group from 15 multiclass data sets in a 9:1 ratio at random, and afterward, these data sets have been combined into 1 training group. The training group has been transferred to the layout to train and learn. Table 7 presents the outcomes of 15 test sets transferred to the model for practically simulating the efficiency of the model applications. It is apparent that the model's accuracy has decreased. It is because data interaction would occur by increasing the number of data resulting in changing the model, and whenever whole data has been combined, there would unavoidably be abnormal points and noises. Due to the fact that such noises and anomalies have not been separated in training, the model's indexes alter, and the robustness decreases.

Table 6. Assessment of characteristics chosen.

Characteristics	Only New Characteristics	12.5% Main Characteristics and New Characteristics	25% Main Characteristics and New Characteristics	37.5% Main Characteristics and New Characteristics	50% Main Characteristics and New Characteristics
Mean accuracy	0.7492	0.9390	0.9350	0.9337	0.9334
Characteristics	62.5% Main Characteristics and New Characteristics	75% Main Characteristics and New Characteristics	87.5% Main Characteristics and New Characteristics	100% Main Characteristics and New Characteristics	
Mean accuracy	0.9335	0.9331	0.9324	0.9353	

Table 7. Layout accuracy on 15 trail sets in the actual simulation.

Data set	Data 1	Data 2	Data 3	Data 4	Data 5	Data 6	Data 7	Data 8
Data number	0.8894	0.8699	0.9097	0.8830	0.9092	0.9096	0.9066	0.9193
Data set	Data 9	Data 10	Data 11	Data 12	Data 13	Data 14	Data 15	Entire
Data number	0.9083	0.9229	0.9241	0.9007	0.9016	0.8966	0.9130	0.9043

Firstly, the efficacy of the features created from the feature construction engineering in the model is determined by sorting the significance of features. Model interpretability can be determined by determining the significance of features. Weight, gain, cover, and so on are general indicators of feature significance. In the XGBoost method [30], the number of times a property appears in a tree has been shown by weight, the mean gain of the slot using the property has been represented by the gain, and the mean coverage of the slot using the property is shown by the cover. According to Figure 4, weight calculates feature significance. The abscissa indicates the names of the beat 45 properties, and the ordinate indicates the assessment score. The origin features are shown by the gray part. The features derived from feature construction engineering are represented by the red mark. It is evident that each of the 16-making properties is in the best 45.

The test trains 15 sets of multiclass classification data sets and tests respectively and uses accuracy as an assessment metric. The accuracy of the trail data sent to the layout before and after optimization based on the main 128 properties is shown in Figure 5. The classification accuracy of the trail group on various layouts with default variables is shown in Figure 5a, and the accuracy of the trail group on the layout applying optimized variables is represented in Figure 5b. For a more intuitive visualization of the variation in accuracy after layouts are optimized, Figure 5a and b are combined, and the mean of the accuracy values for whole sets are adopted, i.e., Figure 5c. Figure 5 shows that the SVM layout with default variables has an accuracy of approximately 0.30, but after optimization, it grows to 0.85, which represents a near 200% advancement. Other models have improved significantly in accuracy after optimization as well. The best accuracy of the suggested AWV model is 0.9217.

Table 3 shows that every data set has about 5000 segments of data; therefore, the CNN layout cannot be used. The semantic relationships among features might also be ignored by several neural networks, such as CNN and long-short-term memory (LSTM) layouts. Thus, in several cases, statistical features according to the manual design could positively affect model accuracy as well. Moreover, the tree-based algorithm outperforms KNN and SVM.

The test set had better performance on the model suggested in this study in comparison to the conventional DML and CNN, as shown in Figure 5.

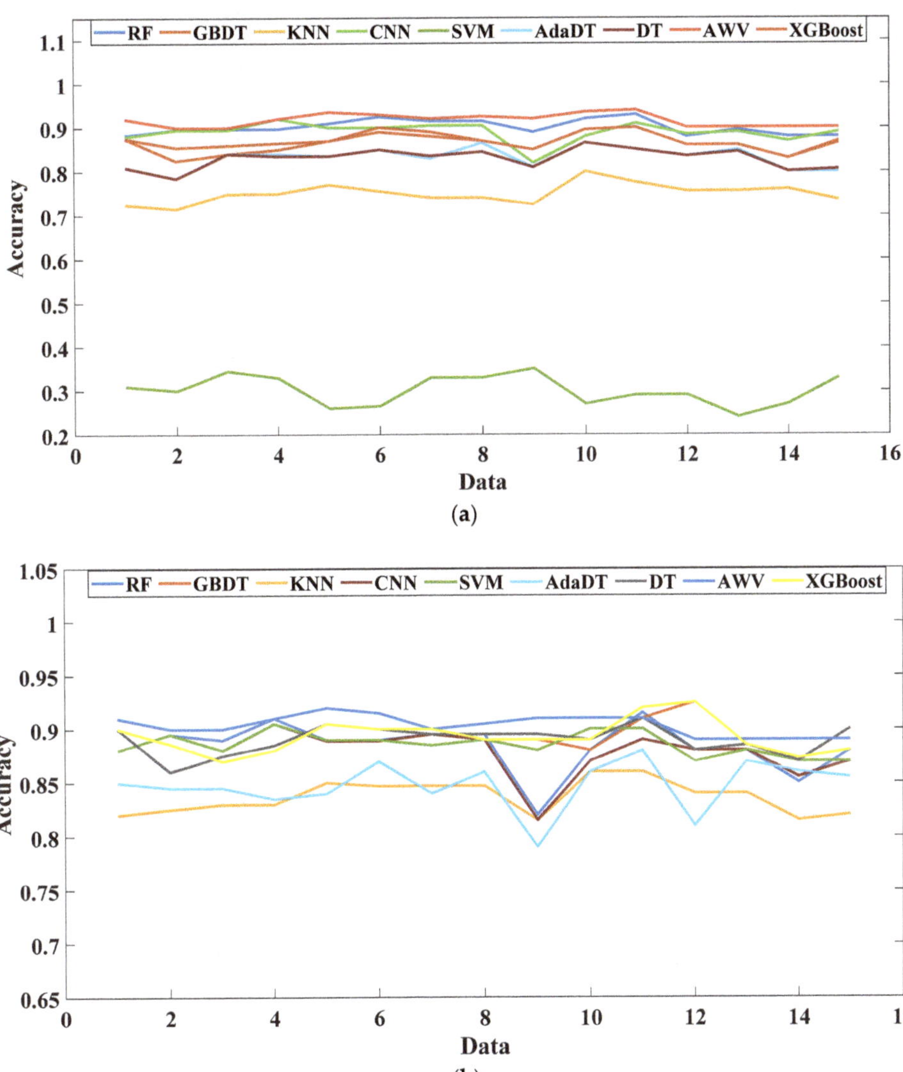

Figure 5. *Cont.*

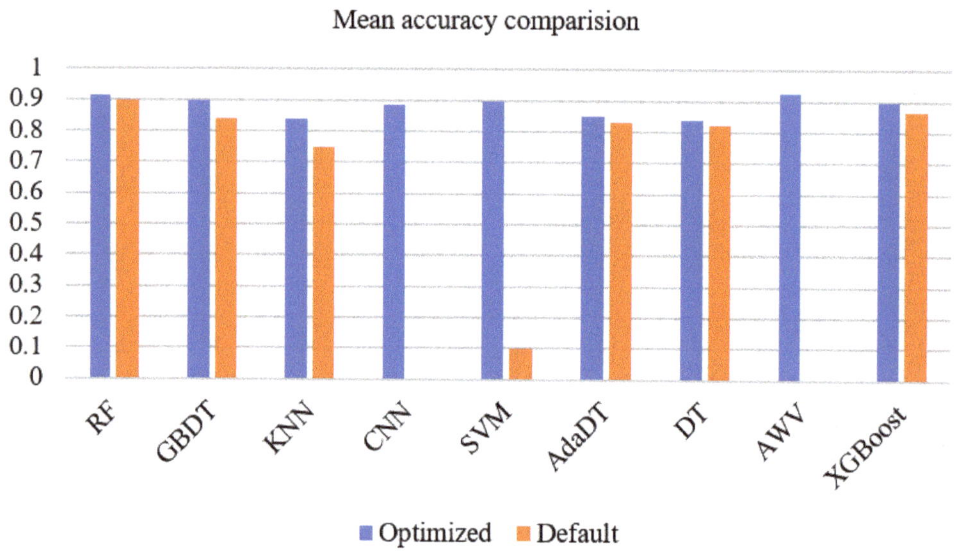

(**c**)

Figure 5. Proficiency comparisons of variables through applying 128 properties (**a**); (**b**) precision over 15 data sets through applying optimum variables; (**c**) mean accuracy comparison.

4. Conclusions

Various SG information as the experimental foundation is used in the present study, and property making for the original data is applied. The layout for identifying faults and cyber-attack in the electrical system is proposed. The present study uses various DML assessment indexes for evaluating the suggested model and conventional DML methods in the experiment. According to the outcomes, the information analyzing process improves the model's accuracy, and the AWV layout detects 37 types of behavior in electrical systems efficiently. As a result, machine learning can be used in the power grid to assist operators in making decisions. In other words, the smart grid operator can always check the health level of the data gathering by the PMUs all around the grid. In the case that any abnormality is detected, the possibility of an intentional cyber-attack exists, and thus, some cautious pre-operation strategies shall be considered to keep the power and demand balance. Moreover, if the data readings from any PMU are unusual, the system operator can decide to estimate the system status without this PMU and rely more on the data coming from the other healthy PMUs.

Author Contributions: Conceptualization, A.A., S.A. and M.A.M.; methodology, A.A., S.A. and M.A.M.; software, A.A., S.A. and M.A.M.; validation, A.A., S.A. and M.A.M.; formal analysis, A.A., S.A. and M.A.M.; investigation, A.A., S.A. and M.A.M.; data curation, A.A., S.A. and M.A.M.; writing—original draft preparation, A.A., S.A. and M.A.M.; writing—review and editing, A.A., S.A. and M.A.M.; visualization, A.A., S.A. and M.A.M.; supervision, A.A., S.A. and M.A.M.; project administration, A.A. and S.A.; funding acquisition, A.A. and S.A. All authors have read and agreed to the published version of the manuscript.

Funding: This research has been funded by the Scientific Research Deanship at the University of Ha'il—Saudi Arabia through project number RG-21079.

Institutional Review Board Statement: Not applicable.

Informed Consent Statement: Not applicable.

Data Availability Statement: Not applicable.

Conflicts of Interest: The authors declare no conflict of interest.

References

1. Almalaq, A.; Albadran, S.; Alghadhban, A.; Jin, T.; Mohamed, M.A. An Effective Hybrid-Energy Framework for Grid Vulnerability Alleviation under Cyber-Stealthy Intrusions. *Mathematics* **2022**, *10*, 2510. [CrossRef]
2. Reich, J.; Schneider, D.; Sorokos, I.; Papadopoulos, Y.; Kelly, T.; Wei, R.; Armengaud, E.; Kaypmaz, C. Engineering of Runtime Safety Monitors for Cyber-Physical Systems with Digital Dependability Identities. In Proceedings of the International Conference on Computer Safety, Reliability, and Security, Lisbon, Portugal, 15 September 2020; Springer: Cham, Switzerland, 2020; pp. 3–17.
3. Li, Y.; Wang, B.; Wang, H.; Ma, F.; Zhang, J.; Ma, H.; Mohamed, M.A. Importance Assessment of Communication Equipment in Cyber-Physical Coupled Distribution Network Based on Dynamic Node Failure Mechanism. *Front. Energy Res.* **2022**, 654. [CrossRef]
4. Zhang, L.; Cheng, L.; Alsokhiry, F.; Mohamed, M.A. A Novel Stochastic Blockchain-Based Energy Management in Smart Cities Using V2S and V2G. *IEEE Trans. Intell. Transp. Syst.* **2022**, 1–8. [CrossRef]
5. Chen, J.; Alnowibet, K.; Annuk, A.; Mohamed, M.A. An effective distributed approach based machine learning for energy negotiation in networked microgrids. *Energy Strategy Rev.* **2021**, *38*, 100760. [CrossRef]
6. Al-Mhiqani, M.N.; Ahmad, R.; Yassin, W.; Hassan, A.; Abidin, Z.Z.; Ali, N.S.; Abdulkareem, K.H. Cyber-security incidents: A review cases in cyber-physical systems. *Int. J. Adv. Comput. Sci. Appl.* **2018**, *1*, 499–508.
7. Luo, Y.; Cheng, L.; Liang, Y.; Fu, J.; Peng, G. Deepnoise: Learning sensor and process noise to detect data integrity attacks in CPS. *China Commun.* **2021**, *18*, 192–209. [CrossRef]
8. Kaouk, M.; Flaus, J.M.; Potet, M.L.; Groz, R. A review of intrusion detection systems for industrial control systems. In Proceedings of the 2019 6th International Conference on Control, Decision and Information Technologies (CoDIT), Paris, France, 23 April 2019; IEEE: Toulouse, France, 2019; pp. 1699–1704.
9. Dehghani, M.; Kavousi-Fard, A.; Dabbaghjamanesh, M.; Avatefipour, O. Deep learning based method for false data injection attack detection in AC smart islands. *IET Gener. Transm. Distrib.* **2020**, *14*, 5756–5765. [CrossRef]
10. Taormina, R.; Galelli, S.; Tippenhauer, N.O.; Salomons, E.; Ostfeld, A.; Eliades, D.G.; Aghashahi, M.; Sundararajan, R.; Pourahmadi, M.; Banks, M.K.; et al. Battle of the attack detection algorithms: Disclosing cyber attacks on water distribution networks. *J. Water Resour. Plan. Manag.* **2018**, *144*, 04018048. [CrossRef]
11. Chang, Q.; Ma, X.; Chen, M.; Gao, X.; Dehghani, M. A deep learning based secured energy management framework within a smart island. *Sustain. Cities Soc.* **2021**, *70*, 102938. [CrossRef]
12. Keshk, M.; Sitnikova, E.; Moustafa, N.; Hu, J.; Khalil, I. An integrated framework for privacy-preserving based anomaly detection for cyber-physical systems. *IEEE Trans. Sustain. Comput.* **2019**, *6*, 66–79. [CrossRef]
13. Huang, Y.; He, T.; Chaudhuri, N.R.; la Porta, T. Preventing Outages under Coordinated Cyber-Physical Attack with Secured PMUs. *IEEE Trans. Smart Grid* **2022**, *13*, 3160–3173. [CrossRef]
14. Alexopoulos, T.A.; Korres, G.N.; Manousakis, N.M. Complementarity reformulations for false data injection attacks on pmu-only state estimation. *Electr. Power Syst. Res.* **2020**, *189*, 106796. [CrossRef]
15. Alexopoulos, T.A.; Manousakis, N.M.; Korres, G.N. Fault location observability using phasor measurements units via semidefinite programming. *IEEE Access* **2016**, *4*, 5187–5195. [CrossRef]
16. Mamuya, Y.D.; Lee, Y.-D.; Shen, J.-W.; Shafiullah, M.; Kuo, C.-C. Application of Machine Learning for Fault Classification and Location in a Radial Distribution Grid. *Appl. Sci.* **2020**, *10*, 4965. [CrossRef]
17. Chaithanya, P.S.; Priyanga, S.; Pravinraj, S.; Sriram, V.S. SSO-IF: An Outlier Detection Approach for Intrusion Detection in SCADA Systems. In *Inventive Communication and Computational Technologies*; Springer: Singapore, 2020; pp. 921–929.
18. Chen, J.; Mohamed, M.A.; Dampage, U.; Rezaei, M.; Salmen, S.H.; Obaid, S.A.; Annuk, A. A multi-layer security scheme for mitigating smart grid vulnerability against faults and cyber-attacks. *Appl. Sci.* **2021**, *11*, 9972. [CrossRef]
19. Avatefipour, O.; Al-Sumaiti, A.S.; El-Sherbeeny, A.M.; Awwad, E.M.; Elmeligy, M.A.; Mohamed, M.A.; Malik, H. An intelligent secured framework for cyberattack detection in electric vehicles' CAN bus using machine learning. *IEEE Access* **2019**, *7*, 127580–127592. [CrossRef]
20. Wang, B.; Ma, F.; Ge, L.; Ma, H.; Wang, H.; Mohamed, M.A. Icing-EdgeNet: A pruning lightweight edge intelligent method of discriminative driving channel for ice thickness of transmission lines. *IEEE Trans. Instrum. Meas.* **2020**, *70*, 1–12. [CrossRef]
21. Alnowibet, K.; Annuk, A.; Dampage, U.; Mohamed, M.A. Effective energy management via false data detection scheme for the interconnected smart energy hub–microgrid system under stochastic framework. *Sustainability* **2021**, *13*, 11836. [CrossRef]
22. Chen, L.; Liu, Z.; Tong, L.; Jiang, Z.; Wang, S.; Dong, J.; Zhou, H. Underwater object detection using Invert Multi-Class Adaboost with deep learning. In Proceedings of the 2020 International Joint Conference on Neural Networks (IJCNN), Glasgow, UK, 19 July 2020; IEEE: Toulouse, France, 2020; pp. 1–8.
23. Shafizadeh-Moghadam, H. Fully component selection: An efficient combination of feature selection and principal component analysis to increase model performance. *Expert Syst. Appl.* **2021**, *186*, 115678. [CrossRef]
24. Roshan, K.; Zafar, A. Deep Learning Approaches for Anomaly and Intrusion Detection in Computer Network: A Review. *Cyber Secur. Digit. Forensics* **2022**, *73*, 551–563.

25. Aceto, G.; Ciuonzo, D.; Montieri, A.; Pescape, A. Traffic classification of mobile apps through multi-classification. In Proceedings of the GLOBECOM 2017-2017 IEEE Global Communications Conference, Singapore, 4 December 2017; IEEE: Toulouse, France, 2017; pp. 1–6.
26. Pham, B.T.; Bui, D.T.; Prakash, I.; Nguyen, L.H.; Dholakia, M.B. A comparative study of sequential minimal optimization-based support vector machines, vote feature intervals, and logistic regression in landslide susceptibility assessment using GIS. *Environ. Earth Sci.* **2017**, *76*, 371. [CrossRef]
27. Jena, M.; Dehuri, S. Decision tree for classification and regression: A state-of-the art review. *Informatica* **2020**, *44*, 405–420. [CrossRef]
28. Chen, R.C.; Caraka, R.E.; Arnita, N.E.; Pomalingo, S.; Rachman, A.; Toharudin, T.; Tai, S.K.; Pardamean, B. An end to end of scalable tree boosting system. *Sylwan* **2020**, *165*, 1–11.
29. Lulli, A.; Oneto, L.; Anguita, D. Mining big data with random forests. *Cogn. Comput.* **2019**, *11*, 294–316. [CrossRef]
30. Franklin, J. The elements of statistical learning: Data mining, inference and prediction. *Math. Intell.* **2005**, *27*, 83–85. [CrossRef]

 *mathematics*

MDPI

Article

Theoretical Bounds on the Number of Tests in Noisy Threshold Group Testing Frameworks

Jin-Taek Seong

Department of Convergence Software, Mokpo National University, Muan 58554, Korea; jtseong@mokpo.ac.kr

Abstract: We consider a variant of group testing (GT) models called noisy threshold group testing (NTGT), in which when there is more than one defective sample in a pool, its test result is positive. We deal with a variant model of GT where, as in the diagnosis of COVID-19 infection, if the virus concentration does not reach a threshold, not only do false positives and false negatives occur, but also unexpected measurement noise can reverse a correct result over the threshold to become incorrect. We aim to determine how many tests are needed to reconstruct a small set of defective samples in this kind of NTGT problem. To this end, we find the necessary and sufficient conditions for the number of tests required in order to reconstruct all defective samples. First, Fano's inequality was used to derive a lower bound on the number of tests needed to meet the necessary condition. Second, an upper bound was found using a MAP decoding method that leads to giving the sufficient condition for reconstructing defective samples in the NTGT problem. As a result, we show that the necessary and sufficient conditions for the successful reconstruction of defective samples in NTGT coincide with each other. In addition, we show a trade-off between the defective rate of the samples and the density of the group matrix which is then used to construct an optimal NTGT framework.

Keywords: noisy threshold group testing; defective samples; number of tests; bounds; COVID-19

MSC: 68Q01

Citation: Seong, J.-T. Theoretical Bounds on the Number of Tests in Noisy Threshold Group Testing Frameworks. *Mathematics* **2022**, *10*, 2508. https://doi.org/10.3390/math10142508

Academic Editors: Gurami Tsitsiashvili and Alexander Bochkov

Received: 9 June 2022
Accepted: 15 July 2022
Published: 19 July 2022

Publisher's Note: MDPI stays neutral with regard to jurisdictional claims in published maps and institutional affiliations.

1. Introduction

Group Testing (GT) is a underdetermined problem in [1], and numerous methods have been developed to solve their problems. GT has become relevant in various problems including probabilistic approaches. The expansion of compressive sensing goes back to the fundamental idea of GT because it is an effort to find sparse signals [2,3]. Recently, academia has begun using the GT method as on vital approach to finding confirmed COVID-19 cases, showing this field's potential importance in these uncertain times [4,5]

The first study for GT was proposed by Dorfman [1]. The background to the emergence of GT is that a large project was conducted in the United States to find soldiers with syphilis during World War II. Syphilis testing of inviduals involves taking a blood sample, then analyzing that to produce a positive or negative result for syphilis in that patient. The syphilis testing carried out at the time was very inefficient since it took a lot of time and money to test all the soldiers one by one [3]. After all, if N soldiers are individually tested for syphilis, N tests are required. Note that the number of soldiers infected with syphilis is very small compared to the total number of soldiers. That is why it is probably inefficient to test every soldier for syphilis one by one, and why the GT technique emerged. The initial GT model was performed in the following way [1]. Several soldiers' blood samples were randomly selected, and the blood was put into a pool and mixed. Then, the blood pool was checked to see if it activated to syphilis or not. A positive result indicates that at least one of the soldiers in the pool was infected with syphilis. A negative result, on the other hand, indicates that all soldiers in the pool were free of syphilis. GT is attractive because the number of tests can be drastically reduced in the case of fewer soldiers infected with syphilis. After these beginnings, GT has mainly been studied with two different approaches, each forming a field of research of its own. One these fields is how to generate GT models. That is, it is a method of selecting samples to be included in one test pool. The second area is to reconstruct defective samples with as few tests as possible. GT loses its benefits if the requirement for a large number of retests leads to as many tests as the number of tests for individual screening.

For GT, various models have been proposed in consideration of how the test results express positive and negative results and the presence or absence of noise. In general, GT's test results told us to see if the pool under being tested contains one or more defective samples. That is, a positive or negative result indicates whether at least one of the defective samples in the pool are present. The model called quantitative GT [3] is a generalization framework of GT. The test result of quantitative GT indicates the number of defective samples in the test pool. There is also another GT model called Threshold Group Testing (TGT) [6]. In the TGT model, a test result of a pool is positive or negative as in conventional GT schemes. However, unlike the conventional GT model, the positive result occurs only when the number of defective samples in the pool is greater than a given threshold. Otherwise, the test outcome is negative. The TGT model is used because it can represent situations in which the test result can be different depending on whether it is high or low, such as the COVID-19 virus concentration. A modified GT model in which measurement noise causes false negatives or false positives is also considered.

TGT problems have been dealt with in various fields such as construction of TGT models [7], theoretical analysis of performance [8], and efficient model design [9,10]. However, there have been no studies so far to quantify how much measurement noise affects performance of TGT models. In this paper, we consider a Noisy Threshold Group Testing (NTGT) model. We provide guidelines for designing a NTGT model that is robust and reliable to measurement noise. To this end, a lower bound on the number of tests is derived using Fano's inequality. We show the trade-off relationship between the sparsity of the group matrix and the defective rate of the signal. And we obtain an upper bound on the probability of an error using the MAP decoding method. We show necessary and sufficient conditions on the number of tests required for finding a set of given defective samples using the lower and upper bounds.

2. Related Work

We look through previous studies and their significance to GT. Then, we will classify each type of problem related to current approaches to GT and consider the issues surrounding these problems. The study of GT first began in 1943 [1]. Dorfman made an effort to find a small number of syphilis-infected soldiers. Dorfman performed the GT with the following procedure. When testing for syphilis, all the soldiers were divided into various groups that were equal in size, then individual testing was only performed on soldiers from the groups that had recoded positive test results. In [1], the optimal group size for a given total number of samples and defective rate was summarized and presented. Later, Sterrett improved the performance by slightly modifying the existing GT method [11]. The main idea of Sterrett's approach is that once the first positive result is obtained, the remaining untested individuals are put in one large grouped and tested. Other than that, there is no difference between Sterrett's method and Dorfman's. If there is a low infection rate, Sterrett's method is more efficient because most of the samples are normal. A more general GT has been presented in [12], in which several algorithms were developed for finding defective samples when no infection rate exists. The paper [12] also provided a link between information theory and GT, introduced a new application of GT, and discussed the generalization of GT.

GT is classified based on types of defective sample distributions and decoding approaches. A probabilistic model uses the assumption that a defective sample is generated from a given probability distribution. On the other hand, the combinatorial model is an attempt to find defective samples without knowledge of probability distributions [13,14]. A typical example of this model is the minmax algorithm [15]. In [16], the results of improved performance in the combinatorial model were presented. Looking at other classes, the adaptive case is a model in which samples to be included in one pool are not independent of the results of previous tests. The samples to be used for the next round are changed each time based on the results of previous tests. Specifically, the method of selecting samples to be included in the next pool is optimized by using the results obtained from previous tests. Conversely, in the non-adaptive model, all tests are performed at the same time by a sample selection process defined in advance. So in this model, every test is independent of each other. This model offers the advantage of being able to test simultaneously regardless of the test order. When predetermined multiple steps are used, the non-adaptive model is extended to multi-stage models [1,17]. In fact, although the adaptive model has more constraints in GT design than the non-adaptive model, the adaptive model generally outperforms the non-adaptive model [3]. However, the recent research in [18] showed limitations in improving the performance of the adaptive model. Non-adaptive GTs are more efficient if all tests are being performed at the same time.

We now look at the significance of certain recent studies on noisy GT. The work in [19] showed the information-theoretic performance of GT with and without measurement noise. Several studies have recently showed interesting and significant performance. In [17], the proposed algorithms uses positive rates in the group to be included for each sample. In this case, if it is greater than the set value, the sample is considered as defective. This approach does not lead to optimal performance in all domains, but it follows a scaling law for a specific domain. In [19], there is separate testing for signals, and all of the group testing is carried out while still considering each sample. That is, although no individual testing is performed, samples use a binary value such as positive or negative. In the case of samples affected by symmetric noise, it was shown that the minimum number of tests reduces to a proportional to $K \log N$ of the optimal information-theoretic bounds for identifying any K defectives samples in a population of N samples [19].

In [20,21], for noisy addition, GT algorithms were presented using message passing and linear programming. Although it does not guarantee optimal performance for decoding complexity, the algorithm proposed in [22] is capable of realistic runtime in terms of that case of a large population. Although many studies have been performed on the noiseless version of GT models, it has been considered as an assumption that the test results are always pure. But this is not realistic. In addition, most of the noisy GT approaches to deal with measurement noise were performed by considering the symmetric noise model such as binary symmetric channel mentioned in channel coding theory. The symmetric noise model referred to in this paper assumes that the test results have the same probability of occurrence of false negatives and false positives. However, asymmetric noise models are more natural than symmetric ones in various applications. For example, data forensics in [23] is an example of using noisy GT models where it identifies to see if recoded files are changed.

3. Noisy Threshold Group Testing Framework

3.1. Problem Statement

We define our NTGT problem. Let be the input $\mathbf{x}$ expressed as a binary vector of size N, $\mathbf{x} = (x_1, x_2, \cdots, x_N)$, $\mathbf{x} \in \{0,1\}^N$. For $i \in [N]$, x_i is the i-th element of $\mathbf{x}$. x_i is expressed in binary to identify either a defective sample or a normal sample. In other words, if the i-th sample is defective, $x_i = 1$, otherwise $x_i = 0$. Throughout this work, we assume that x_i has the following probability,

$$\Pr(x_i = \alpha) = \begin{cases} 1 - \delta & \text{if } \alpha = 0, \\ \delta & \text{if } \alpha = 1, \end{cases} \tag{1}$$

where δ is the defective sample rate, and α is a dummy variable. In this case, the defective sample rate is less than 0.5, $0 < \delta < 0.5$, which is considered a small value for GT problems.

As mentioned earlier, one of the key points in the GT problems is to determine which samples to participate in a pool. In this paper, samples to be included in the pool are selected using a non-adaptive model. We use a matrix as a more concise way to define the samples to be included in the pool. Let be the group matrix which has M rows and N columns as denoted $\mathbf{A} \in \{0,1\}^{M \times N}$, where M is the number of tests in the NTGT model. Note that we aim for a small M as the number of tests required to reconstruct the signal $\mathbf{x}$. When the j-th test includes i-th sample x_i and performs GT, it is expressed as $A_{ji} = 1$. Otherwise, $A_{ji} = 0$. Whether i-th sample is included in the j-th test and performs GT, is expressed as a binary value, i.e., 0 or 1, of each element A_{ji} of the group matrix. Although the d-Separable matrix and the d-Disjunct matrix [3] were used to design the group matrix, the approach of randomly selecting the elements of the group matrix is also known to be a good design method [3]. For $i \in [N]$ and $j \in [M]$, A_{ji} is identically independent distributed as follows:

$$\Pr\left(A_{ji} = \alpha\right) = \begin{cases} 1 - \gamma & \text{if } \alpha = 0, \\ \gamma & \text{if } \alpha = 1, \end{cases} \tag{2}$$

where γ denotes the sparsity of the group matrix and the range of γ is $0 < \gamma < 1$. As γ increases, the density of the group matrix also increases. Conversely, as they get smaller, increasingly sparse group matrices are designed. It should be noted that the computational complexity of the GT framework also increases when a group matrix is constructed from a large γ. Therefore, it is necessary to design GT frameworks with as low as possible the sparsity of group matrices while improving the reconstruction performance. We will consider how the relationship between δ and γ affects the number of tests for signal reconstruction.

The reason we are considering the NTGT model is as follows. Consider a model that could be used for the diagnosis of COVID-19 infection. There are cases in which the COVID-19 test showed false positive or false negative results when the concentration of the virus was low or contaminated. The current diagnosis of COVID-19 infection is positive when the virus concentration is above a certain level. During the incubation period or early stage of infection, the virus concentration is low, and false negative results may be obtained. In addition, even if the COVID-19 infection is confirmed using a precise and accurate diagnostic method, the result is sometimes reversed due to unexpected measurement noise. Throughout this work, a NTGT model suited to these challenges is considered. In other words, we consider the best approach to a TGT scheme where positive and negative cases occur by the quatitative concentration, and we consider an additive noise model because measurement noise can reverse the results. In a recent study [24], for the diagnosis of COVID-19 infection, false positives and false negatives were reported to be between 0.1% and 4.5%, respectively. Next, we obtain lower and upper performance bounds on the NTGT model in Sections 4 and 5.

TGT is different from conventional GT models. In conventional GT, if at least one defective sample exists in one test, the output is positive without measurement noise. However, TGT is positive when there is a number of defective samples greater than the predefined threshold T. For example, $T = 3$ means that a positive result occurs only when there are at least three defective samples in the pool. Once there is only one defective sample in the pool, its result would be negative. In other words, the result in the pool becomes positive only when it is above T for TGT models, also whether it is negative or positive in the diagnosis of COVID-19 infection depends on whether the virus concentration is high or low. The conventional GT uses $T = 1$. The following (3) presents an output for a TGT model. Let z_j be the result of the j-th test pool, which does not suffer from noise, where $z_j = 1$ indicates a positive result and 0 for a negative result, $j \in [M]$, $\mathbf{z} = (z_1, z_2, \cdots, z_M)$.

$$z_j = \begin{cases} 0 & \text{if } \sum_{i=1}^{N} A_{ji} x_i < T, \\ 1 & \text{if } \sum_{i=1}^{N} A_{ji} x_i \geq T, \end{cases} \tag{3}$$

Through this paper, we consider the NTGT framework with measurement noise. Assume a model whose results can be flipped due to the measurement noise. z_j is the pure result of the pool test, and its result converts from positive to negative and vice versa due to additive noise. For the NTGT model, the additive noise is defined as follows:

$$\Pr\left(e_j = \alpha\right) = \begin{cases} 1 - \eta & \text{if } \alpha = 0, \\ \eta & \text{if } \alpha = 1, \end{cases} \tag{4}$$

where η is the measurement noise, and we assume all e_j are independent of each other. Therefore, the j-th output y_j in the NTGT model can be written as

$$y_j = z_j \oplus e_j \tag{5}$$

where the symbol $\oplus$ denotes the logical operation XOR. We denote $\mathbf{y} = (y_1, y_2, \cdots, y_M)$ and $\mathbf{e} = (e_1, e_2, \cdots, e_M)$.

Figure 1 shows an example of this NTGT. In this example, two samples out of ten are defective, which is realized from (1). As shown in Figure 1, the number of tests is 7, $M = 7$. The 7×10 group matrix is constructed by (2) mentioned above. For noiseless version, the vector $\mathbf{z}$ is $(0, 0, 1, 0, 0, 0, 0)$ with $T = 2$. In the third test only, the number of defective samples becomes two, and the test result is positive. When additive noise is added as defined in (4), the output is $\mathbf{y} = (1, 0, 1, 0, 0, 0, 0)$.

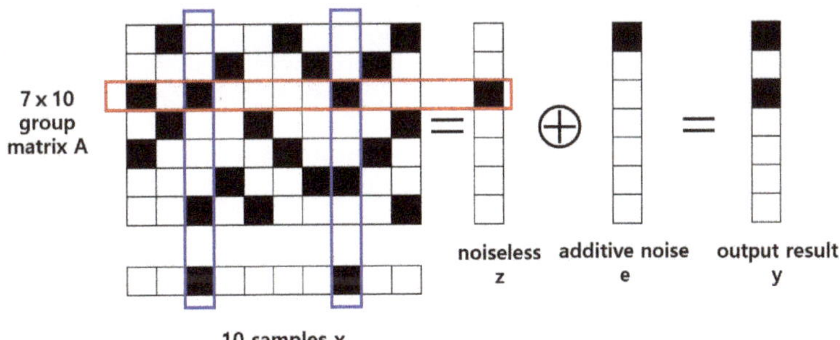

Figure 1. One example of NTGT where $M = 7$, $N = 10$, $T = 2$, the black boxes denotes 1 s, and white ones 0 s.

3.2. Decoding

We use a maximum a posteriori (MAP) method to reconstruct a signal $\mathbf{x}$ in the NTGT.

$$\hat{\mathbf{x}} = \arg\max_{\mathbf{x}} P(\mathbf{x}|\mathbf{y}, \mathbf{A}) \tag{6}$$

The posteriori probability in (6) is as follows:

$$\begin{aligned} P(\mathbf{x}|\mathbf{y}, \mathbf{A}) &= \frac{P(\mathbf{x}, \mathbf{y}, \mathbf{A})}{P(\mathbf{y}, \mathbf{A})} \\ &\propto P(\mathbf{x}, \mathbf{y}, \mathbf{A}) \\ &= \sum_{\mathbf{e}} P(\mathbf{x}, \mathbf{y}, \mathbf{A}, \mathbf{e}) \\ &= \sum_{\mathbf{e}} P(\mathbf{x}) P(\mathbf{A}) P(\mathbf{e}) P(\mathbf{y}|\mathbf{x}, \mathbf{A}, \mathbf{e}) \end{aligned} \tag{7}$$

The last line of (7) is obtained using independent conditions, while the conditional probability $P(\mathbf{y}|\mathbf{x}, \mathbf{A}, \mathbf{e})$ is an indicator function that satisfies the following condition:

$$P(\mathbf{y}|\mathbf{x}, \mathbf{A}, \mathbf{e}) = \begin{cases} 1 & \text{if } \mathbf{y} = \mathbf{z} \oplus \mathbf{e}, \\ 0 & \text{if } \mathbf{y} \neq \mathbf{z} \oplus \mathbf{e}, \end{cases} \tag{8}$$

We define an error event if $\hat{\mathbf{x}}$ from (6) is not the same as the true realization of $\mathbf{x}$. In other words, the probability of an error is expressed as $P_E = \Pr\{\hat{\mathbf{x}} \neq \mathbf{x}\}$.

3.3. Bounds for Group Testing Schemes

Now consider the number of tests on successful decoding in the conventional GT models. The number of tests required to identify K defective samples out of all N samples for an adaptive GT algorithm with perfect reconstruction denotes as $m(N, K)$. Moreover, for the case of a non-adaptive model, the number of tests is defined as $\bar{m}(N, K)$. The number of tests N required for individual testing is greater than $\bar{m}(N, K)$. Adaptive GT models require less or equal number of tests than those of non-adaptive GTs because they check the results of previous tests and perform the next tests, $m(N, K) \leq \bar{m}(N, K)$. Even if the number of defective samples is one, at least one test must be performed, $1 \leq m(N, K)$. Therefore, the number of tests has a wide range as follows:

$$1 \leq m(N, K) \leq \bar{m}(N, K) \leq N \tag{9}$$

From an information-theoretic bound, the minimum number of tests M for a GT framework with a sample space is obtained as [3],

$$M \geq \log_2 |\mathcal{S}| \tag{10}$$

where $\mathcal{S}$ denotes the sample space. In addition, an information-theoretic performance is presented even for a GT framework with small error probability. It is expressed as an upper bound of the error probability for the number of tests required for successful decoding. This GT algorithm performs in such a way as the following bound on successful probability P_s for decoding of defective samples [25]:

$$P_s \leq \frac{M}{\log_2 \binom{N}{K}} \tag{11}$$

In the past half century, many studies on GT models have been performed, and among them, well-known and important GT algorithms are introduced next. The first one to be considered is the binary splitting algorithm [3]. This algorithm solves the existing GT problems efficiently and is applicable to the adaptive GT models. So far, the reason this algorithm is used for GT problems is because of its simplicity and good performance. The number of tests required to reconstruct defective samples using the binary splitting algorithm is known through the following bounds:

$$M = \begin{cases} N & \text{if } N \leq 2K - 2, \\ (\log_2 \sigma + 2)K + p - 1 & \text{if } N \geq 2K - 1, \end{cases} \tag{12}$$

where σ is the number of samples to be included for one test, and p is a uniquely determined nonnegative integer conditioning $p < K$.

Next, the definite defectives algorithm [26] is considered. This algorithm is suitable for non-adaptive GT models because an unknown input signal can be reconstructed using all of the test results at the same time through an iterative process. The feature of the definite defectives algorithm is attractive in that it can eliminate false negatives that may occur during the reconstruction process. As a result, the use of the definite defectives algorithm is more useful in applications where false negatives are sensitive or should not be present. For given N and K, the definite defective algorithm has the following lower bound for the number of tests M required for identifying defective samples if it is allowed an error rate of σ,

$$M \geq (1 - \sigma) \log_2 \binom{N}{K} \tag{13}$$

This can be observed that (11) and (13) coincide with the same in the perfect reconstruction of defective samples.

4. Necessary Condition for Complete Recovery

4.1. Lower Bound

In this section, we take into account a necessary condition for the number of tests required to identify defective samples in the NTGT model. We obtain the necessary condition using Fano's inequality theorem [27] presented in information theory. Fano's inequality is mainly exploited in channel coding theory, and describes the connection between error probability and entropy. In addition, in [28], the authors reviewed GT problems comprehensively and in-depth from an information theory perspective. The lower bound on the probability of an error is obtained by considering Fano's inequality theorem. From this lower bound, we are lead to the necessary condition for the number of tests to find all defective samples for the NTGT model. We first explain Fano's inequality theorem before deriving the necessary condition.

Theorem 1 (Fano's inequality [27])**.** *Suppose there are random variables A and B of finite size. If the decoding function Φ that finds A by considering B is used, the following inequality holds:*

$$1 + P(\Phi(B) \neq A) \log_2 |A| \geq H(A|B) \tag{14}$$

where $P(\Phi(B) \neq A)$ is the probability of an error for the decoding function Φ, and the conditional entropy $H(A|B)$ is defined as follows:

$$H(A|B) = -\sum_{\alpha \in A} \sum_{\beta \in B} P_{AB}(\alpha, \beta) \log P_{A|B}(\alpha|\beta) \tag{15}$$

where P_{AB} and $P_{A|B}$ are the joint probability and conditional probability, respectively.

In the NTGT problem, we are able to obtain a lower bound on the probability of an error. This lower bound shows the minimum number of tests required to reconstruct an unknown signal, regardless of which decoding function is used. In this paper, our lower bound is a variant of the results obtained in [8]. Compared to [8], this work obtains the lower bound taking into account the measurement noise. However, the overall procedure of derivation is similar to each other because it uses Fano's inequality theorem.

Theorem 2 (Lower bound)**.** *For any decoding function with the unknown sample signal defined in (1) and the measurement noise defined in (4), a necessary condition for the probability of error P_E to be less than an arbitrary small and positive value ρ for $P_E < \rho$ holds such that*

$$\frac{NH(\delta) - M + MH(\eta) - 1}{N} < \rho \tag{16}$$

where $H(\cdot)$ is the entropy function.

Proof of Theorem 2. Let $\hat{\mathbf{x}}$ be the estimated signal of $\mathbf{x}$ found using the decoding function. Considering the following process in terms of a Markov chain, we can say $\mathbf{x} \rightarrow (\mathbf{y}, \mathbf{A}) \rightarrow \hat{\mathbf{x}}$. Then, the following inequality is satisfied,

$$H(\mathbf{x}|\mathbf{y}, \mathbf{A}) \leq H(\mathbf{x}|\hat{\mathbf{x}}) \tag{17}$$

Further, from Fano's inequality described in (14), the conditional entropy is bounded by

$$H(\mathbf{x}|\mathbf{y}, \mathbf{A}) \leq 1 + P_E \log_2 \left(2^N - 1\right) \tag{18}$$

Then, the probability of error is bounded in terms of the conditional entropy and the total number of samples N,

$$P_E \geq \frac{H(\mathbf{x}|\mathbf{y}, \mathbf{A}) - 1}{N} \tag{19}$$

It needs to tackle the conditional entropy $H(\mathbf{x}|\mathbf{y}, \mathbf{A})$. Let us divide and expand the following conditional entropy in more detail:

$$\begin{aligned} H(\mathbf{x}|\mathbf{y}, \mathbf{A}) &= H(\mathbf{x}) - I(\mathbf{x}; \mathbf{y}, \mathbf{A}) \\ &= H(\mathbf{x}) - (I(\mathbf{x}; \mathbf{A}) + I(\mathbf{x}; \mathbf{y}|\mathbf{A})) \\ &\overset{(a)}{=} H(\mathbf{x}) - (H(\mathbf{y}|\mathbf{A}) - H(\mathbf{y}|\mathbf{A}, \mathbf{x})) \end{aligned} \tag{20}$$

where $I(\cdot)$ is mutual information, and equality (a) comes from the fact that $\mathbf{x}$ and $\mathbf{A}$ are independent of each other. Note that the smaller the term on the right side of (19), the lower the minimum value of the probability of error. This means that the conditional entropy, $H(\mathbf{x}|\mathbf{y}, \mathbf{A})$, should be small as possible. As a result, on the last line of the right side in (20), the conditional entropy $H(\mathbf{y}|\mathbf{A})$ should be large; conversely, the conditional entropy $H(\mathbf{y}|\mathbf{A}, \mathbf{x})$ should be small.

To do this, let us find the maximum and minimum values of the two conditional entropies, respectively.

$$\begin{aligned} H(\mathbf{y}|\mathbf{A}) \leq H(\mathbf{y}) &= H(\mathbf{z} \oplus \mathbf{e}) \\ &\leq M \end{aligned} \tag{21}$$

where the first inequality is due to the definition of conditional entropy, and the last inequality comes from the fact that the result y_j is either 0 or 1, y_j values are independent of each other, and the maximum binary entropy is 1 in the case that $\Pr(y_j = 0) = \Pr(y_j = 1)$. Next, we take into account the other conditional entropy $H(\mathbf{y}|\mathbf{A}, \mathbf{x})$ which is minimized,

$$\begin{aligned} H(\mathbf{y}|\mathbf{A}, \mathbf{x}) &= H(\mathbf{z} \oplus \mathbf{e}|\mathbf{A}, \mathbf{x}) \\ &= H(\mathbf{e}) \\ &= MH(\eta) \end{aligned} \tag{22}$$

where the second equality comes from how the randomness of $\mathbf{z}$ vanishes if $\mathbf{x}$ and $\mathbf{A}$ are known, the last equality being due to the independent events of $\mathbf{e}$. Using (21) and (22), (20) can be rewritten as

$$H(\mathbf{x}|\mathbf{y}, \mathbf{A}) \leq NH(\delta) - M + MH(\eta) \tag{23}$$

Finally, if (19) is changed to satisfy the condition $P_E < \rho$ where ρ is a small, positive value and $\rho > 0$, the following condition holds:

$$\frac{NH(\delta) - M + MH(\eta) - 1}{N} < \rho \tag{24}$$

This completes the proof of Theorem 2. $\square$

4.2. Construction of Noisy Threshold Group Testing

We now consider the result obtained from Theorem 2. First, Theorem 2 can be expressed as the ratio of the number of tests to the total number of samples as follows:

$$\frac{M}{N} > \frac{H(\delta) - \rho}{1 - H(\eta)} \tag{25}$$

It is advantageous to use the NTGT framework until the point N and M are equal. Otherwise, when $M > N$, individual testing becomes more effective than GT. This shows that NTGT can theoretically be used under the following noise conditions:

$$H(\eta) < 1 + \rho - H(\delta) \tag{26}$$

To design an NTGT framework, how to construct a group matrix is important. The key to this is shown in the proof of Theorem 2. Looking carefully at the conditions under which the inequality of conditional entropy holds in (21), the maximum conditional entropy $H(\mathbf{y}|\mathbf{A})$ is obtained when the following conditions are satisfied: $\Pr(y_j = 0) = \Pr(y_j = 1)$. This means that the NTGT system should be designed so that the output has an equal probability of being 0 or 1. Since $\mathbf{x}$ and $\mathbf{A}$ are independent of each other, the probability of an output of 0 is as follows:

$$\Pr(y_j = 0) = \sum_{t=0}^{T-1} \binom{N}{t} (\delta\gamma)^t (1 - \delta\gamma)^{N-t} = \frac{1}{2} \tag{27}$$

As shown in (27), it can be seen that there is a trade-off between δ and γ. In other words, to reconstruct a sparse signal, a high-density group matrix needs to be generated and used. Conversely, if the signal is not sparse, the group matrix should be designed with low density.

5. Sufficient Condition for Average Performance

5.1. Upper Bound

Now we prove there is an upper bound on the probability of errors from the MAP decoding used in NTGT. We divide the proof of the upper bound into two parts: one considers the definition of the error event and the other part formulates the probability of errors.

We rewrite the a posteriori probability.

$$P(\mathbf{x}|\mathbf{y}, \mathbf{A}) \propto \sum_{\mathbf{e}} P(\mathbf{x}) P(\mathbf{A}) P(\mathbf{e}) 1_{\mathbf{y}=\mathbf{z}\oplus\mathbf{e}} \tag{28}$$

Note that both $\mathbf{A}$ and $\mathbf{y}$ are given and known. Using MAP decoding, we estimate with (28)

$$\hat{\mathbf{x}} = \arg\max_{\mathbf{x}} \sum_{\mathbf{e}} P(\mathbf{x}) P(\mathbf{A}) P(\mathbf{e}) 1_{\mathbf{y}=\mathbf{z}\oplus\mathbf{e}} \tag{29}$$

An error event occurs if there is a feasible vector $\bar{\mathbf{x}} \neq \mathbf{x}$, such that

$$\sum_{\mathbf{v}} P(\bar{\mathbf{x}}) P(\mathbf{v}) 1_{\mathbf{y}=\mathbf{w}\oplus\mathbf{v}} \geq \sum_{\mathbf{e}} P(\mathbf{x}) P(\mathbf{e}) 1_{\mathbf{y}=\mathbf{z}\oplus\mathbf{e}} \tag{30}$$

where $\mathbf{w} = \sum_{A_{ji}\bar{x}_i \geq T} \mathbf{A}\bar{\mathbf{x}}$ comes from (3), and $\mathbf{v}$ comes from a realization from (4). When given $\mathbf{y}$, $\mathbf{A}$, and $\mathbf{x}$, we have one vector $\mathbf{e}$, such that $\mathbf{e} = \mathbf{z} \oplus \mathbf{y}$. Then we can rewrite (30).

$$P(\bar{\mathbf{x}}) P_{\mathbf{v}}(\mathbf{y} \oplus \mathbf{w}) \geq P(\mathbf{x}) P_{\mathbf{e}}(\mathbf{y} \oplus \mathbf{z}) \tag{31}$$

Therefore, an error event becomes equivalent to there existing a pair $(\bar{\mathbf{x}}, \mathbf{v})$ such that

$$\begin{aligned} \bar{\mathbf{x}} &\neq \mathbf{x}, \\ \mathbf{y} &= \mathbf{w} \oplus \mathbf{v} = \mathbf{z} \oplus \mathbf{e}, \\ P(\bar{\mathbf{x}}) P_{\mathbf{v}}(\mathbf{y} \oplus \mathbf{w}) &\geq P(\mathbf{x}) P_{\mathbf{e}}(\mathbf{y} \oplus \mathbf{z}) \end{aligned} \tag{32}$$

So far, we have defined the error event and now we will derive an upper bound on the probability of error. When given $\mathbf{x}$ and $\mathbf{e}$, we let $P(\mathcal{I}|\mathbf{x}, \mathbf{e})$ be the conditional error probability. We have an average error probability as follows:

$$P_E = \sum_{\mathbf{x}} \sum_{\mathbf{e}} P(\mathbf{x}, \mathbf{e}) P(\mathcal{I}|\mathbf{x}, \mathbf{e}) \tag{33}$$

We now introduce two typical sets that were defined in [27] (Ch.3.1). Let $\mathcal{A}_{[\mathbf{e}]\varepsilon}^M$ and $\mathcal{A}_{[\mathbf{x}]\varepsilon}^N$ be typical sets of $\mathbf{x}$ and $\mathbf{e}$ with respect to $P(\mathbf{x})$ and $P(\mathbf{e})$ as defined in (1) and (4).

For any positive number ε and sufficiently large numbers of N and M, the two typical sets are defined as

$$\mathcal{A}_{[x]\varepsilon}^{N} = \left\{ \mathbf{x} \in 2^{N} : \left| -\frac{1}{N} \log P(\mathbf{x}) - H(\delta) \right| \leq \varepsilon \right\} \tag{34}$$

and

$$\mathcal{A}_{[e]\varepsilon}^{M} = \left\{ \mathbf{e} \in 2^{M} : \left| -\frac{1}{M} \log P(\mathbf{e}) - H(\eta) \right| \leq \varepsilon \right\} \tag{35}$$

From the Shannon–McMillan–Breiman theorem [27] (Ch.16.8), we obtain the following two bounds:

$$P\left(\left| -\frac{1}{N} \log P(\mathbf{x}) - H(\delta) \right| \leq \varepsilon \right) \geq 1 - \varepsilon \tag{36}$$

and

$$P\left(\left| -\frac{1}{M} \log P(\mathbf{e}) - H(\eta) \right| \leq \varepsilon \right) \geq 1 - \varepsilon \tag{37}$$

Now we define the space of the pair $(\mathbf{x}, \mathbf{e})$ with respect to the two typical sets. Let $\mathcal{U}$ and $\mathcal{U}^{c}$ be the sets for the pair $(\mathbf{x}, \mathbf{e})$ such that

$$\mathcal{U} = \left\{ \mathbf{x} \in 2^{N}, \mathbf{e} \in 2^{M} : \left(\mathbf{x} \in \mathcal{A}_{[x]\varepsilon}^{N} \bigcap \mathbf{e} \in \mathcal{A}_{[e]\varepsilon}^{M} \right) \right\} \tag{38}$$

and

$$\mathcal{U}^{c} = \left\{ \mathbf{x} \in 2^{N}, \mathbf{e} \in 2^{M} : \left(\mathbf{x} \notin \mathcal{A}_{[x]\varepsilon}^{N} \bigcup \mathbf{e} \notin \mathcal{A}_{[e]\varepsilon}^{M} \right) \right\} \tag{39}$$

where $\mathcal{U}$ is the joint typical set for the pair $(\mathbf{x}, \mathbf{e})$, since $\mathbf{x}$ and $\mathbf{e}$ are independent.

Theorem 3 (Upper bound). *In NTGT one, a distribution of defective samples defined in* (1) *and noise probability defined in* (4)*, for any small ε, the ratio of the number of tests M to the total number of samples N is upper-bounded:*

$$\frac{M}{N} > \frac{H(\delta) + \varepsilon}{1 - H(\eta) - \varepsilon} \tag{40}$$

Proof of Theorem 3. The probability of error is bounded as

$$
\begin{aligned}
P_{E} &= \sum_{(\mathbf{x},\mathbf{e}) \in \mathcal{U}} P(\mathbf{x})P(\mathbf{e})P(\mathcal{I}|\mathbf{x},\mathbf{e}) + \sum_{(\mathbf{x},\mathbf{e}) \in \mathcal{U}^{c}} P(\mathbf{x})P(\mathbf{e})P(\mathcal{I}|\mathbf{x},\mathbf{e}) \\
&\overset{(a)}{\leq} \sum_{(\mathbf{x},\mathbf{e}) \in \mathcal{U}} P(\mathbf{x})P(\mathbf{e})P(\mathcal{I}|\mathbf{x},\mathbf{e}) + \sum_{(\mathbf{x},\mathbf{e}) \in \mathcal{U}^{c}} P(\mathbf{x})P(\mathbf{e}) \\
&\overset{(b)}{\leq} \sum_{(\mathbf{x},\mathbf{e}) \in \mathcal{U}} P(\mathbf{x})P(\mathbf{e})P(\mathcal{I}|\mathbf{x},\mathbf{e}) + 2\varepsilon
\end{aligned}
\tag{41}
$$

where (a) is due to $P(\mathcal{I}|\mathbf{x},\mathbf{e}) \leq 1$, and (b) comes from the following,

$$
\begin{aligned}
\sum_{(\mathbf{x},\mathbf{e}) \in \mathcal{U}^{c}} P(\mathbf{x})P(\mathbf{e}) &= 1 - \sum_{(\mathbf{x},\mathbf{e}) \in \mathcal{U}} P(\mathbf{x})P(\mathbf{e}) \\
&= 1 - \sum_{\mathbf{x} \notin \mathcal{A}_{[x]\varepsilon}^{N}} P(\mathbf{x}) \sum_{\mathbf{e} \notin \mathcal{A}_{[e]\varepsilon}^{M}} P(\mathbf{e}) \\
&\leq 1 - (1 - \varepsilon)(1 - \varepsilon) \\
&\leq 2\varepsilon
\end{aligned}
\tag{42}
$$

This is because $\mathbf{A}$ is randomly generated as defined in (2); then we can define the following event as

$$\mathcal{E}(\mathbf{x}, \mathbf{e}; \bar{\mathbf{x}}, \mathbf{v}) = \left\{ (\mathbf{x}, \mathbf{e}; \bar{\mathbf{x}}, \mathbf{v}) : \mathbf{z} \oplus \mathbf{e} = \mathbf{w} \oplus \mathbf{v} \right\} \tag{43}$$

The conditional error probability $P(\mathcal{I}|\mathbf{x}, \mathbf{e})$ is the probability of the union of all the events in (43) with respect to all pairs $(\bar{\mathbf{x}}, \mathbf{v})$ that satisfy (32). Thus, the conditional error probability in (33) can be rewritten as

$$P(\mathcal{I}|\mathbf{x}, \mathbf{e}) = \Pr\left\{ \bigcup_{\bar{\mathbf{x}}, \mathbf{v}: P(\bar{\mathbf{x}})P(\mathbf{v}) \geq P(\mathbf{x})P(\mathbf{e})} \mathcal{E}(\mathbf{x}, \mathbf{e}; \bar{\mathbf{x}}, \mathbf{v}) \right\} \tag{44}$$

Using the union bound in (41), we have the following bound:

$$
\begin{aligned}
P_E &\leq \sum_{(\mathbf{x},\mathbf{e})\in\mathcal{U}} P(\mathbf{x})P(\mathbf{e}) \sum_{(\bar{\mathbf{x}},\mathbf{v}): P(\bar{\mathbf{x}})P(\mathbf{v})\geq P(\mathbf{x})P(\mathbf{e})} P\big(\mathcal{E}(\mathbf{x}, \mathbf{e}; \bar{\mathbf{x}}, \mathbf{v})\big) + 2\varepsilon \\
&= \sum_{(\mathbf{x},\mathbf{e})\in\mathcal{U}} P(\mathbf{x})P(\mathbf{e}) \sum_{(\bar{\mathbf{x}},\mathbf{v})} P\big(\mathcal{E}(\mathbf{x}, \mathbf{e}; \bar{\mathbf{x}}, \mathbf{v})\big)\Phi(\mathbf{x}, \bar{\mathbf{x}}, \mathbf{e}, \mathbf{v}) + 2\varepsilon
\end{aligned}
\tag{45}
$$

where $\Phi(\mathbf{x}, \bar{\mathbf{x}}, \mathbf{e}, \mathbf{v})$ is the indicator function, such that $P(\bar{\mathbf{x}})P(\mathbf{v}) \geq P(\mathbf{x})P(\mathbf{e})$.

$$\Phi(\mathbf{x}, \bar{\mathbf{x}}, \mathbf{e}, \mathbf{v}) = \begin{cases} 1 & \text{if } P(\bar{\mathbf{x}})P(\mathbf{v}) \geq P(\mathbf{x})P(\mathbf{e}) \\ 0 & \text{if } P(\bar{\mathbf{x}})P(\mathbf{v}) < P(\mathbf{x})P(\mathbf{e}) \end{cases} \tag{46}$$

The indicator function is bounded [29] (Ch. 5.6) for $0 < s \leq 1$.

$$\Phi(\mathbf{x}, \bar{\mathbf{x}}, \mathbf{e}, \mathbf{v}) \leq \left(\frac{P(\bar{\mathbf{x}})P(\mathbf{v})}{P(\mathbf{x})P(\mathbf{e})} \right)^s \tag{47}$$

For $s = 1$ in (47), we have the following bound:

$$P_E \leq \sum_{(\mathbf{x},\mathbf{e})\in\mathcal{U}} \sum_{(\bar{\mathbf{x}},\mathbf{v})} P(\bar{\mathbf{x}})P(\mathbf{v})P\big(\mathcal{E}(\mathbf{x}, \mathbf{e}; \bar{\mathbf{x}}, \mathbf{v})\big) + 2\varepsilon \tag{48}$$

From the definition in (43), note that the probability $P\big(\mathcal{E}(\mathbf{x}, \mathbf{e}; \bar{\mathbf{x}}, \mathbf{v})\big)$ is

$$P\big(\mathcal{E}(\mathbf{x}, \mathbf{e}; \bar{\mathbf{x}}, \mathbf{v})\big) = \Pr(\mathbf{w} \oplus \mathbf{v} = \mathbf{z} \oplus \mathbf{e}) \tag{49}$$

where

$$
\begin{aligned}
P_E &\leq \sum_{(\mathbf{x},\mathbf{e})\in\mathcal{U}} \sum_{(\bar{\mathbf{x}},\mathbf{v})} P(\bar{\mathbf{x}})P(\mathbf{v})P\big(\mathcal{E}(\mathbf{x}, \mathbf{e}; \bar{\mathbf{x}}, \mathbf{v})\big) + 2\varepsilon \\
&= \sum_{(\mathbf{x},\mathbf{e})\in\mathcal{U}, \|\mathbf{e}\oplus\mathbf{v}\|_0=d_2} \sum_{\|\bar{\mathbf{x}}\|_0=d_1} \sum_{\|\mathbf{e}\oplus\mathbf{v}\|_0=d_2} P(\bar{\mathbf{x}})P(\mathbf{v})P\Big(\mathbf{z} \oplus \mathbf{w} = \mathbf{e} \oplus \mathbf{v}\Big| \|\bar{\mathbf{x}}\|_0 = d_1, \|\mathbf{e} \oplus \mathbf{v}\|_0 = d_2\Big)
\end{aligned}
\tag{50}
$$

In (50), we find the following probability depending on the number of nonzero elements d_1 and d_2:

$$
\begin{aligned}
P\Big(\mathbf{z} \oplus \mathbf{w} = \mathbf{e} \oplus \mathbf{v}\Big| \|\bar{\mathbf{x}}\|_0 = d_1, \|\mathbf{e} \oplus \mathbf{v}\|_0 = d_2\Big) &= \prod_{j=1}^{M} P\Big(z_j \oplus w_j = e_j \oplus v_j\Big| \|\bar{\mathbf{x}}\|_0 = d_1, \|\mathbf{e} \oplus \mathbf{v}\|_0 = d_2\Big) \\
&= P\Big(z_j \oplus w_j = 1\Big| \|\bar{\mathbf{x}}\|_0 = d_1, \|\mathbf{e} \oplus \mathbf{v}\|_0 = d_2\Big)^{d_2} \\
&\quad \times P\Big(z_j \oplus w_j = 0\Big| \|\bar{\mathbf{x}}\|_0 = d_1, \|\mathbf{e} \oplus \mathbf{v}\|_0 = d_2\Big)^{M-d_2} \\
&= (1 - P_0)^{d_2} P_0^{M-d_2}
\end{aligned}
\tag{51}
$$

where each row is independent. Given this, we define the following probability:

$$P_0 \triangleq \Pr\big(z_j \oplus w_j = 0\big| \|\bar{\mathbf{x}}\|_0 = d_1\big) \tag{52}$$

We can divide P_0 in (52) into two parts. If $d_1 < T$,

$$
\begin{aligned}
P_0 &= \Pr(z_j = 0)\Pr(w_j = 0) + \Pr(z_j = 1)\Pr(w_j = 1) \\
&= \Pr(z_j = 0)
\end{aligned}
\tag{53}
$$

Otherwise,

$$
\begin{aligned}
P_0 &= \Pr(z_j = 0)\left(\sum_{t=0}^{T-1}\binom{d_1}{t}\gamma^t(1-\gamma)^{(d_1-t)}\right) + \Pr(z_j = 1)\left(\sum_{t=T}^{d_1}\binom{d_1}{t}\gamma^t(1-\gamma)^{(d_1-t)}\right) \\
&= P_{z,0}(\delta,\gamma)P_{w,0}(d_1,\gamma) + (1 - P_{z,0}(\delta,\gamma))(1 - P_{w,0}(d_1,\gamma))
\end{aligned}
\tag{54}
$$

where

$$
\begin{aligned}
P_{z,0}(\delta,\gamma) &\triangleq \Pr(z_j = 0) = \sum_{t=0}^{T-1}\binom{N}{t}(\delta\gamma)^t(1-\delta\gamma)^{N-t}, \\
P_{w,0}(d_1,\gamma) &\triangleq \Pr(w_j = 0) = \sum_{t=0}^{T-1}\binom{d_1}{t}\gamma^t(1-\gamma)^{d_1-t}
\end{aligned}
\tag{55}
$$

The maximum for P_0 by looking at $P_{z,0}(\delta,\gamma) = 1/2$ and $P_{w,0}(d_1,\gamma) = 1/2$ from the fact that P_0 in (54) is concave with respect to $P_{z,0}(\delta,\gamma)$ and $P_{w,0}(d_1,\gamma)$. Therefore, its bound is

$$
P_0 \leq \frac{1}{2}
\tag{56}
$$

Using (51) and (56), (50) can be bounded as follows:

$$
\begin{aligned}
P_E &\leq 2^{-M}\sum_{d_1=0,x\neq\bar{x}}\sum_{(x,e)\in\mathcal{U},\bar{x}:\|\bar{x}\|_0=d_1}P(\bar{x})\left(\sum_{\mathbf{v}}P(\mathbf{v})\right) + 2\varepsilon \\
&\leq 2^{-M}\sum_{d_1=0,x\neq\bar{x}}\sum_{(x,e)\in\mathcal{U},\bar{x}:\|\bar{x}\|_0=d_1}P(\bar{x}) + 2\varepsilon \\
&\leq 2^{-M}\sum_{x\in\mathcal{A}^N_{[x]\varepsilon}}\sum_{e\in\mathcal{A}^M_{[e]\varepsilon}}\sum_{d_1=0,x\neq\bar{x}}P(\bar{x}) + 2\varepsilon \\
&= 2^{-M}\left|\mathcal{A}^N_{[x]\varepsilon}\right|\cdot\left|\mathcal{A}^M_{[e]\varepsilon}\right|\sum_{d_1=0,x\neq\bar{x}}P(\bar{x}) + 2\varepsilon \\
&\leq 2^{-M}\left|\mathcal{A}^N_{[x]\varepsilon}\right|\cdot\left|\mathcal{A}^M_{[e]\varepsilon}\right| + 2\varepsilon \\
&\leq 2^{-M}2^{N(H(\delta)+\varepsilon)}2^{M(H(\eta)+\varepsilon)} + 2\varepsilon \\
&= 2^{N(H(\delta)+\varepsilon)+M(H(\eta)+\varepsilon)-M} + 2\varepsilon
\end{aligned}
\tag{57}
$$

As the probability of error is less than 1, the exponent term on the right side of (57) is bounded by

$$
N(H(\delta)+\varepsilon) + M(H(\eta)+\varepsilon) - M < 0
\tag{58}
$$

Then, the ratio of M to N is

$$
\frac{M}{N} > \frac{H(\delta)+\varepsilon}{1 - H(\eta) - \varepsilon}
\tag{59}
$$

This completes the proof of Theorem 3. $\quad\square$

5.2. Discussion for Necessary and Sufficient Conditions

In this section, we discuss the results obtained from Theorems 2 and 3. The result from Theorem 2 allows us to solve the lower bound in the NTGT problem using Fano's inequality. The minimum number of tests required to recover all defective samples with

δ probability out of N samples is also obtained. In other words, Theorem 2 is a necessary condition for any probability of error to be smaller than ρ. Conversely, Theorem 3 leads to the upper bound on the probability of an error using the MAP decoding method. This condition refers to the upper bound on performance and is the sufficient condition to allow us to reconstruct defective samples.

We show that the results of Theorems 2 and 3 coincide with each other. Finding and presenting the necessary and sufficient conditions for the number of tests required in the NTGT problem is significant for TGT. In addition, as shown in (27) above, a system design method for NTGT was proposed so that the probability that a test result is 0 and the probability that it is 1 are the same depending on threshold T.

6. Conclusions

In this paper, we considered a NTGT problem where the test result is positive when the number of defective samples in a pool equals or greater than a certain threshold. Recently, when performing GT for the diagnosis COVID-19 infection, if the sample's virus concentration did not sufficiently reach the threshold, false positives or false negatives can occur, so in this work we dealt with this TGT framework. In addition, a noise model was added in case pure results were flipped due to unexpected measurement noise. We took into account how many tests were needed to successfully reconstruct a small defective sample with the NTGT problem. To this end, we aimed to find the necessary and sufficient conditions for the number of tests required. For the necessary condition, we obtained the lower bound on the number of tests using Fano's inequality theorem. Next, the upper bound on performance defined by the probability of error was derived using the MAP decoding method. This result leads to the sufficient condition for identifying all defective samples in the NTGT problem. In this paper, we have shown that the necessary and sufficient conditions are consistent with the NTGT framework. In addition, we presented that the relationship between the defective rate of the input signal and the sparsity of the group matrix should be considered to design an optimal NTGT system.

Funding: National Research Foundation of Korea: NRF-2020R1I1A3071739.

Institutional Review Board Statement: Not applicable.

Informed Consent Statement: Not applicable.

Data Availability Statement: Not applicable

Conflicts of Interest: The authors declare no conflict of interest.

References

1. Dorfman, R. The Detection of Defective Members of Large Populations. *Ann. Math. Stat.* **1943**, *14*, 436–440. [CrossRef]
2. Donoho, D.L. Compressed Sensing. *IEEE Trans. Inf. Theory* **2006**, *52*, 1289–1306. [CrossRef]
3. Du, D.-Z.; Hwang, F.-K. *Pooling Designs and Nonadaptive Group Testing: Important Tools for DNA Sequencing*; World Scientific: Singapore, 2006.
4. Verdun, C.M.; Fuchs, T.; Harar, P.; Elbrächter, D.; Fischer, D.S.; Berner, J.; Grohs, P.; Theis, F.J.; Krahmer, F. Group Testing for SARS-CoV-2 Allows for Up to 10-Fold Efficiency Increas Across Realistic Scenarios and Testing Strategies. *Front. Public Health* **2021**, *9*, 583377. [CrossRef] [PubMed]
5. Mutesa, L.; Ndishimye, P.; Butera, Y.; Souopgui, J.; Uwineza, A.; Rutayisire, R.; Ndoricimpaye, E.L.; Musoni, E.; Rujeni, N.; Nyatanyi, T.; et al. A pooled testing strategy for identifying SARS-CoV-2 at low prevalence. *Nature* **2021**, *589*, 276–280. [CrossRef] [PubMed]
6. Damaschke, P. Threshold group testing. *Gen. Theory Inf. Transf. Comb. LNCS* **2006**, *4123*, 707–718.
7. Bui, T.V.; Kuribayashi, M.; Cheraghchi, M.; Echizen, I. Efficiently Decodable Non-Adaptive Threshold Group Testing. *IEEE Trans. Inf. Theory* **2019**, *65*, 5519–5528. [CrossRef]
8. Seong, J.-T. Theoretical Bounds on Performance in Threshold Group Testing. *Mathematics* **2020**, *8*, 637. [CrossRef]
9. Chen, H.; Bonis, A.D. An almost optimal algorithm for generalized threshold group testing with inhibitors. *J. Comput. Biol.* **2011**, *18*, 851–864. [CrossRef] [PubMed]
10. De Marco, G.; Jurdzinski, T.; Rozanski, M.; Stachowiak, G. Subquadratic non-adaptive threshold group testing. *Fundam. Comput. Theory* **2017**, *111*, 177–189.

11. Sterrett, A. On the Detection of Defective Members of Large Populations. *Ann. Math. Stat.* **1957**, *28*, 1033–1036. [CrossRef]
12. Sobel, M.; Groll, P.A. Group testing to eliminate efficiently all defectives in a binomial sample. *Bell Syst. Tech. J.* **1959**, *38*, 1179–1252. [CrossRef]
13. Allemann, A. An Efficient Algorithm for Combinatorial Group Testing. In *Information Theory, Combinatorics, and Search Theory*; Lecture Notes in Computer Science; Springer: Berlin/Heidelberg, Germany, 2013; Volume 7777.
14. Srivastava, J.N. *A Survey of Combinatorial Theory*; North Holland Publishing Co.: Amsterdam, The Netherlands, 1973.
15. Riccio, L.; Colbourn, C.J. Sharper bounds in adaptive group testing. *Taiwan. J. Math.* **2000**, *4*, 669–673. [CrossRef]
16. Leu, M.-G. A note on the Hu–Hwang–Wang conjecture for group testing. *ANZIAM J.* **2008**, *49*, 561–571. [CrossRef]
17. Chan, C.L.; Che, P.H.; Jaggi, S.; Saligrama, V. Non-adaptive probabilistic group testing with noisy measurements: Near-optimal bounds with efficient algorithms. In Proceedings of the 49th Annual Allerton Conference on Communication, Control, and Computing, Monticello, IL, USA, 28–30 September 2011.
18. Atia, G.K.; Saligrama, V. Boolean Compressed Sensing and Noisy Group Testing. *IEEE Trans. Inf. Theory* **2012**, *58*, 1880–1901. [CrossRef]
19. Malyutov, M. The separating property of random matrices. *Math. Notes Acad. Sci. USSR* **1978**, *23*, 84–91. [CrossRef]
20. Sejdinovic, D.; Johnson, O. Note on noisy group testing: Asymptotic bounds and belief propagation reconstruction. In Proceedings of the 48th Annual Allerton Conference on Communication, Control, and Computing, Monticello, IL, USA, 29 September–1 October 2010.
21. Malioutov, D.; Malyutov, M. Boolean compressed sensing: LP relaxation for group testing. In Proceedings of the IEEE International Conference on Acoustics, Speech and Signal Processing (ICASSP), Kyoto, Japan, 25–30 March 2012.
22. Bondorf, S.; Chen, B.; Scarlett, J.; Yu, H.; Zhao, Y. Sublinear-time non-adaptive group testing with O(k log n) tests via bit-mixing coding. *IEEE Trans. Inf. Theory* **2021**, *67*, 1559–1570. [CrossRef]
23. Goodrich, M.T.; Atallah, M.J.; Tamassia, R. Indexing information for data forensics. In Proceedings of the Third International Conference on Applied Cryptography and Network Security, New York, NY, USA, 7–10 June 2005.
24. Mistry, D.A.; Wang, J.Y.; Moeser, M.E.; Starkey, T.; Lee, L.Y. A systematic review of the sensitivity and specificity of lateral flow devices in the detection of SARS-CoV-2. *BMC Infect. Dis.* **2021**, *21*, 828. [CrossRef] [PubMed]
25. Baldassini, L.; Johnson, O.; Aldridge, M. The capacity of adaptive group testing. In Proceedings of the IEEE International Symposium on Information Theory, Istanbul, Turkey, 7–12 July 2013.
26. Aldridge, M.; Baldassini, L.; Johnson, O. Group Testing Algorithms: Bounds and Simulations. *IEEE Trans. Inf. Theory* **2014**, *60*, 3671–3687. [CrossRef]
27. Cover, T.M.; Thomas, J.A. *Elements of Information Theory*; Wiley: Hoboken, NJ, USA, 2009.
28. Aldridge, M.; Johnson, O.; Scarlett, J. Group Testing: An Information Theory Perspective, *Found. Trends Commun. Inf. Theory* **2019**, *15*, 196–392. [CrossRef]
29. Gallager, R. *Information Theory and Reliable Communication*; John Wiley and Sons: Hoboken, NJ, USA, 1968.

Article

Limiting Distributions of a Non-Homogeneous Markov System in a Stochastic Environment in Continuous Time

P. -C. G. Vassiliou

Department of Statistical Sciences, University College London, Gower St., London WC1E 6BT, UK; vasiliou@math.auth.gr

Abstract: The stochastic process non-homogeneous Markov system in a stochastic environment in continuous time (S-NHMSC) is introduced in the present paper. The ordinary non-homogeneous Markov process is a very special case of an S-NHMSC. I studied the expected population structure of the S-NHMSC, the first central classical problem of finding the conditions under which the asymptotic behavior of the expected population structure exists and the second central problem of finding which expected relative population structures are possible limiting ones, provided that the limiting vector of input probabilities into the population is controlled. Finally, the rate of convergence was studied.

Keywords: non-homogeneous Markov systems; asymptotic behavior; rate of convergence

MSC: 60J10; 60J20; 60J15

1. Introductory Notes

Citation: Vassiliou, P.-C.G. Limiting Distributions of a Non-Homogeneous Markov System in a Stochastic Environment in Continuous Time. *Mathematics* **2022**, *10*, 1214. https://doi.org/10.3390/math10081214

Academic Editors: Gurami Tsitsiashvili and Alexander Bochkov

Received: 10 March 2022
Accepted: 5 April 2022
Published: 7 April 2022

Publisher's Note: MDPI stays neutral with regard to jurisdictional claims in published maps and institutional affiliations.

The stochastic process of a non-homogeneous Markov system in a stochastic environment (S-NHMS) in discrete time was introduced in [1]. The main goal was to satisfy the need for a more realistic stochastic model in populations with various entities, which were possible to be categorized in a finite number of exhaustive and exclusive states. The expected population structure is studied, that is, the distribution of the expected number of memberships in each state. Note that in the population, apart from the transition of memberships among the states, there are transitions to the external environment, often called wastage from the population, and flow of memberships in the population (system) in the various states, often called recruitment.

The S-NHMS in discrete time is a generalization of the stochastic concept of an NHMS in discrete time, which incorporated the idea of having a pool $\Im_I(t)$ of transition probability matrices to choose from, the roots of which were in [2,3], for the special case where the transition matrices are Leslie matrices.

The stochastic process of an NHMS was first introduced in [4]. This new concept provided a more general framework for a number of Markov chain models in manpower systems, which was actually the initial motive. For examples, see [5–10].

There are also a large number and a great diversity of applied probability models that could be accommodated in this general framework. A simple fact that shows the dynamics of the concept of an NHMS is, as we will show later, that the well known simple Markov chain is a very special case of an NHMS.

In the present paper, we study the development of a continuous time version of a S-NHMS. The choice in practice between a stochastic process in discrete and continuous time is partly a matter of realism and partly one of convenience. With regard to realism, for example, usually one would want to deal with the transitions between the states of the members of the population in continuous time. However, in practice, the computational advantages of discrete time, as well as the mental process of the researcher, leads all too often to the choice of a discrete time process. On the other hand, continuous time models are often more amenable to mathematical analysis and this may count many times in their favor. Having developed both versions of the theory of S-NHMS, more choices are at our disposal, and hence, a more complete version of the entire theory.

A first concise and complete presentation of the theory of non-homogeneous Markov process exists in [11], Section 8.9. There, apart from building a rigorous foundation of the subject, in the respective references, one could also find the initial founders of the subject. Reference [12] started a period of intense study of non-homogeneous Markov processes. Strong ergodicity for continuous time non-homogeneous Markov processes, using mean visit times, was studied in [13]. Important results on the strong ergodicity for continuous time non-homogeneous Markov processes, using criteria on the functions of intensity transition matrices, were provided by [14–16]. I will make extensive use of these results in the present paper.

The estimates of rate of convergence for non-homogeneous Markov processes were studied in a series of papers [17–21]. For Markov systems in continuous time results, could be found in [22–25].

Estimations of the transition intensities in NHMS in continuous time were provided by [26] for various cases of missing data. In [27], transition intensities were studied for homogeneous Markov systems (HMS) in continuous time, as well as the relation between the volume of the attainable expected population structures at time t and the trace and rank of the intensity matrix.

In [28], the authors studied, for closed HMS in continuous time, the stability of size order of elements in an expected population structure as $t \to \infty$. The state sizes of the elements of the expected population structures and their distributions for an HMS in continuous time were studied in [29] with the use of factorial moments. In [30], the author discussed the case of closed HMS with finite capacities of the states. In [31], the close relation between $M/M/k/T/T$ queues and close HMS in continuous time is presented. More recent results on NHMS in continuous time could be found in [32], while a more recent review on the subject was given by [33].

The paper is organized as follows: In Section 2, I define in detail for the first time the stochastic process S-NHMSC. I also show that the ordinary non-homogeneous Markov process is a special case of an S-NHMSC. Furthermore, I clarify that the open homogeneous Markov models and the ordinary NHMS in continuous time are special cases of the S-NHMSC.

In Section 3, I evaluate the expected population structure of the S-NHMSC at any time t, as a function of the basic parameters of the population by establishing the appropriate differential and integral equation it satisfies.

In Section 4, I study the central classical problem, that of finding the conditions under which the asymptotic behavior of the expected population structure $\mathbb{E}[\mathbf{N}(t)]$ as $t \to \infty$ exists, and finding its limit in closed analytic form as a function of the limits of the basic parameters of the system. The second central problem is finding which expected relative population structures are possible limiting ones, provided that we control the limiting vector of input probabilities into the population. We prove that the set $\mathcal{A}^{\infty}$ of asymptotically expected relative population structure $\mathbb{E}[\mathbf{q}(t)]$, under asymptotic input control of the S-NHMSC, is a convex hull of the points, which are functions of the left eigenvector of a certain limiting transition probability matrix and the limiting transition intensity matrices of the inherent non-homogeneous Markov process.

I conclude this section by studying an important question, which logically arises, that is, what is the rate of convergence to asymptotically attainable structures in an S-NHMSC. In fact, I am interested in finding conditions under which the rate is exponential, because then, the practical value of the asymptotic result is greater.

Finally, in Section 5, I present an illustrative example from manpower planning.

2. The S-NHMS in Continuous Time

I will start by presenting the concept of a non-homogeneous Markov system in a stochastic environment in continuous time (S-NHMSC). Let $\{T(t), t \geq 0\}$, a known continuous function of time or a realization of a known stochastic process denoting the total number of members in the system. Let $\mathbb{S} = \{1, 2, ..., k\}$ be the set of states that are assumed to be exclusive and exhaustive. The state of the system at any time t is represented by the expected population structure:

$$\mathbb{E}[\mathbf{N}(t)] = [\mathbb{E}[N_1(t)], \mathbb{E}[N_2(t)], ..., \mathbb{E}[N_k(t)]],$$

where $\mathbb{E}[N_i(t)]$ is the expected number of members of the population at time t. Another representation of the state of the system is provided by the relative expected population structure:

$$\mathbb{E}[\mathbf{q}(t)] = \frac{\mathbb{E}[\mathbf{N}(t)]}{T(t)} = [\mathbb{E}[q_1(t)], \mathbb{E}[q_2(t)], ..., \mathbb{E}[q_k(t)]].$$

Furthermore, among the states of the system, as in the case of a non-homogeneous Markov process ([11]), at the infinitesimal time interval $[t, t + \delta t)$, the probabilities of members of the system to move from state i to state j are generated by the transition intensities $r_{ij}(t)$:

$$p_{ij}(t, t + \delta t) = r_{ij}(t)\delta t + o(\delta t), \text{ for } i \neq j \in \mathbb{S}. \tag{1}$$

It is important to note at this point that (1) is valid as long as during the interval $[t, t + \delta t)$ the transition intensities $r_{ij}(t)$ will operate. When taking a step up the ladder towards reality, I will assume a stochastic mechanism of selecting the values of $r_{ij}(t)$, and the equation will be altered accordingly.

Furthermore, let state $k + 1$ represent members leaving the population and assume that $r_{i,k+1}(t)$ is the transition intensity for a member of the population in state i to leave in the time interval $[t, t + \delta t)$:

$$p_{ik+1}(t, t + \delta t) = r_{ik+1}(t)\delta t + o(\delta t), \text{ for } i \neq j \in \mathbb{S}. \tag{2}$$

The transition intensities $r_{jj}(t)$ are defined by:

$$r_{jj}(t) = -\sum_{\substack{i=1 \\ i \neq j}}^{k+1} r_{ji}(t) \text{ for } j \in \mathbb{S}. \tag{3}$$

Let $\mathbf{R}(t) = \{r_{ij}(t)\}_{i,j \in S}$ be the matrix of transition intensities at time t and $\mathbf{r}_{k+1}^{\top}(t) = [r_{1,k+1}(t), r_{2,k+1}(t), ..., r_{k,k+1}(t)]^{\top}$ be the vector of leaving intensities at time t. Now, let $p_{0i}(t, t + \delta t)$ be the probability of a new member to enter the population in state i, given that it will enter the population in the time interval $[t, t + \delta t)$ and let $\mathbf{p}_0(t, t + \delta t) = [p_{01}(t, t + \delta t), p_{02}(t, t + \delta t), ..., p_{0k}(t, t + \delta t)]$. Define the following probabilities:

$$\hat{p}_{ij}(t, t + \delta t) = p_{ij}(t, t + \delta t) + p_{ik+1}(t, t + \delta t)p_{0j}(t, t + \delta t)$$

$$= r_{ij}(t)\delta t + r_{i,k+1}(t)\delta t p_{0j}(t, t + \delta t) + o(\delta t). \tag{4}$$

Now, let:

$$q_{ij}(t) = \lim_{\delta t \to 0} \frac{\hat{p}_{ij}(t, t + \delta t)}{\delta t} = r_{ij}(t) + r_{i,k+1}(t)p_{0j}(t) \text{ for } i \neq j, \tag{5}$$

be the transition intensity of a membership to move to state j in the time interval $[t, t + \delta t)$, given that it was in state i at time t. To visualize this deeper, let there be $T(t)$ memberships at the beginning of the interval $[t, t + \delta t)$, and each member of the population holds one. During the interval $[t, t + \delta t)$, members are leaving the population and at the exit they give their memberships to their replacement, who is distributed among the states with probabilities $\mathbf{p}_0(t, t + \delta t)$ at the end of the interval. Furthermore, let $\mathbf{Q}(t) = \{q_{ij}(t)\}_{i,j \in \mathbb{S}}$ be the matrix of transition intensities of the memberships. Assume that $\mathbf{Q}(t)$ is measurable and that $\sup_{ij \in \mathbb{S}}\{|q_{ij}(t)|\}$ is integrable on every finite interval of t. We call the Markov process defined by the matrix of intensities $\{\mathbf{Q}(t), t \geq 0\}$ the imbedded or inherent Markov process of the S-NHMSC.

Assume now that in the infinitesimal time interval $[t, t + \delta t)$, the system has the choice of selecting a transition intensity matrix from the pool:

$$\mathcal{R}_I(t) = \{\mathbf{R}_1(t), \mathbf{R}_2(t), ..., \mathbf{R}_\nu(t)\}, \tag{6}$$

such that $\mathbf{R}_i(t)\mathbf{1}^\top + \mathbf{r}_{k+1}^\top(t) = \mathbf{0}$ for $i = 1, 2, ..., \nu$ and for every t. Furthermore, assume that it makes its choice in a stochastic way, and more specifically, in the infinitesimal time interval $[t, t + \delta t)$, the probability of selecting an intensity matrix from the set $\mathcal{R}_I(t)$ is given by

$$c_{ij}(t, t + \delta t) = \mathbb{P}\{\mathbf{R}(t + \delta t) = \mathbf{R}_j(t + \delta t) \mid \mathbf{R}(t) = \mathbf{R}_j(t)\}$$

$$= z_{ij}(t)\delta t + o(\delta t) \text{ for } i \neq j, t \geq 0, \tag{7}$$

and $z_{ii}(t)$ is defined to be:

$$z_{ii}(t) = -\sum_{j \neq i} z_{ij}(t), \ i, j \in I, t \geq 0,$$

and let $c_i(0)$ for $i = 1, 2, ..., k$ be the probabilities of the initial states.

Let $\mathbf{Z}(t) = \{z_{ij}(t)\}_{i,j \in I}$ be the above intensity matrix and assume that $\mathbf{Z}(t)$ is measurable for every $t \geq 0$ and that $\sup_{i \in I}\{|z_{ii}(t)|\}$ is integrable on every finite interval of time. Then, the intensity matrices $\{\mathbf{Z}(t)\}_{t \geq 0}$ define a non-homogeneous Markov process, which we call the compromise non-homogeneous Markov process of the S-NHMSC. The word 'compromise' is selected in the sense that it is the outcome of the choice of strategy under the various pressures in the environment. We call a process like the one described above a non-homogeneous Markov system in a stochastic environment in continuous time (S-NHMSC).

We defined the S-NHMSC in the most general way, in order to provide an inclusive framework that could accommodate a large variety of applied probability models. Furthermore, in the following, some basic questions will be answered within this general framework. However, it is of great importance, in order to increase our intuition about the potential power of applicability of the present theory and in order to place it at the right position in the pyramid of progress towards reality, to make the following comments. Firstly, when:

$$T(t) = 1, \mathbf{p}_{k+1}(t) = 0, \mathbf{p}_0(t) = 0 \text{ for every } t > 0 \text{ and}$$

$$\mathcal{R}_I(t) = \{\mathbf{R}(t)\} \text{ for every } t > 0,$$

then the *S-NHMSC* is the ordinary non-homogeneous Markov process, which has found applications in almost all areas.

Secondly, when:

$$\mathbf{p}_{k+1}(t) = \mathbf{p}_{k+1}, \mathbf{p}_0(t) = \mathbf{p}_0, \mathcal{R}_I(t) = \{\mathbf{R}\} \text{ for every } t > 0,$$

then the *S-NHMSC* is the open homogeneous Markov model applied extensively in manpower systems (see [5,34]).

Thirdly, when:

$$\mathcal{R}_I(t) = \{\mathbf{R}(t)\} \text{ for every } t > 0,$$

then the S-NHMSC is the ordinary NHMS in continuous time, which is a general framework for many applied probability models (see [35,36]).

3. The Expected Population Structure of the S-NHMSC

We will now study the problem of finding the expected population structure $\mathbb{E}[\mathbf{N}(t)]$ in terms of the basic functions of the parameters of the system. We call basic functions of the parameters the least number of parameters that uniquely determine an S-NHMSC. These are the functions $\{\mathcal{R}_I(t)\}_{t\geq 0}$, $\{\mathbf{Z}(t)\}_{t\geq 0}$, $\{T(t)\}_{t\geq 0}$, $\{\mathbf{r}_{k+1}(t)\}_{t\geq 0}$, $\{\mathbf{p}_0(t)\}_{t\geq 0}$, the initial population structure $\mathbf{N}(0)$, and the initial probabilities $c_j(0)$. These are defined by:

$$D(t) = dT(t)/dt \ \text{ or } \ T(t+\delta t) - T(t) = D(t)\delta t + o(\delta t). \tag{8}$$

Let $N_0(t, t+\delta t)$ be the random variable which represents the number of new members entering the population in the infinitesimal time interval $[t, t+\delta t)$. Then, since the number of losses from the population is a random variable, with the distribution for each state $i \in \mathbb{S}$, the binomial $\mathcal{B}(N_i(t), r_{i,k+1}(t)\delta t)$ conditional on $N_i(t)$, we have:

$$\mathbb{E}[N_0(t, t+\delta t)] = \sum_{i=1}^{k} \mathbb{E}[N_i(t)]r_{i,k+1}(t)\delta t + D(t)\delta t. \tag{9}$$

Furthermore, let $N_{ij}(t, t+\delta t)$ be the random variable representing the number of members of the system moving from state i to state j in the time interval $[t, t+\delta t)$. Then, these flows from i to $j \in \mathbb{S}$ are multinomial random variables, in the sense that:

$$\mathbb{E}\big[N_{ij}(t, t+\delta t)\big] = \mathbb{E}\big[\mathbb{E}\big[N_{ij}(t, t+\delta t) \mid N_i(t), \mathcal{R}_I(t)\big]\big]$$

$$= \mathbb{E}[N_i(t)]\mathbb{E}\big[p_{ij}(t, t+\delta t)\big]$$

$$= \mathbb{E}[N_i(t)]\mathbb{E}\big[r_{ij}(t)\big]\delta t + o(\delta t) \text{ for } i \neq j \in \mathbb{S}, \tag{10}$$

and:

$$\mathbb{E}[N_{ii}(t, t+\delta t)] = \mathbb{E}[\mathbb{E}[N_{ii}(t, t+\delta t) \mid N_i(t), \mathcal{R}_I(t)]]$$

$$= \mathbb{E}[N_i(t)]\mathbb{E}[p_{ii}(t, t+\delta t)]$$

$$= \mathbb{E}[N_i(t)] + \mathbb{E}[N_i(t)]\mathbb{E}[r_{ii}(t)]\delta t + o(\delta t) \text{ for } i \neq j \in \mathbb{S}. \tag{11}$$

Consequently, we have:

$$\mathbb{E}\big[N_j(t+\delta t)\big] = \sum_{i \neq j} \mathbb{E}[N_i(t)]\big[\mathbb{E}\big[r_{ij}(t)\big]\delta t + r_{i,k+1}(t)p_{0j}(t, t+\delta t)\big]$$

$$+\mathbb{E}\big[N_j(t)\big]\Big[1 + \mathbb{E}\big[r_{jj}(t)\big]\delta t + r_{jk+1}(t)\delta t p_{0j}(t, t+\delta t)\Big]$$

$$+ D(t)\delta t p_{0j}(t, t+\delta t) + o(\delta t). \tag{12}$$

Equation (12), for all $j \in \mathbb{S}$, could be written in matrix notation:

$$\frac{d\mathbb{E}[\mathbf{N}(t)]}{dt} = \mathbb{E}[\mathbf{N}(t)]\mathbb{E}[\mathbf{Q}(t)] + D(t)\mathbf{p}_0(t), \tag{13}$$

where:

$$\mathbb{E}[\mathbf{Q}(t)] = \mathbb{E}[\mathbf{R}(t)] + \mathbf{r}_{k+1}^{\top}(t)\mathbf{p}_0(t). \tag{14}$$

We will now prove that the sum of the rows of the matrix $\mathbb{E}[\mathbf{Q}(t)]$ is equal to zero. We have:

$$\mathbb{E}[\mathbf{Q}(t)]\mathbf{1}^{\top} = \mathbb{E}[\mathbf{R}(t)]\mathbf{1}^{\top} + \mathbf{r}_{k+1}^{\top}(t)\mathbf{p}_0(t)\mathbf{1}^{\top}$$

$$= \sum_{j=1}^{\nu} \mathbb{P}\big[\mathbf{R}(t) = \mathbf{R}_j(t)\big]\mathbf{R}_j(t)\mathbf{1}^{\top} + \mathbf{r}_{k+1}^{\top}(t)$$

$$= -\sum_{j=1}^{\nu} \mathbb{P}\big[\mathbf{R}(t) = \mathbf{R}_j(t)\big]\mathbf{r}_{k+1}^{\top}(t) + \mathbf{r}_{k+1}^{\top}(t)$$

$$= -\mathbf{r}_{k+1}^{\top}(t) + \mathbf{r}_{k+1}^{\top}(t) = 0. \tag{15}$$

Hence, the matrix $\mathbb{E}[\mathbf{Q}(t)]$ is an intensity matrix and defines a non-homogeneous Markov process which, by analogy with the ordinary NHMS in discrete time [5,35], we call the expected embedded or inherent non-homogeneous Markov process for the S-NHMSC. Assume that $\int_0^t \mathbb{E}[\mathbf{Q}(u)]du < \infty$ for all $t \geq 0$, then there exists a unique transition function (see [36] paragraph 8.9) $\mathbb{E}\big[\mathbf{P}_q(.,.)\big]$, such that:

$$\lim_{h+h' \to 0} \frac{\mathbb{E}\Big[\mathbf{P}_q\big(t - h, t + h'\big)\Big] - \mathbf{I}}{h + h'} = \mathbb{E}[\mathbf{Q}(t)], \tag{16}$$

for all $t \notin E$, where $E \subset [0, \infty)$ is a set of Lebesgue measure zero. Moreover, $\mathbb{E}\big[\mathbf{P}_q(.,.)\big]$ satisfies the integral matrix equations:

$$\mathbb{E}\big[\mathbf{P}_q(s,t)\big] = \mathbf{I} + \int_s^t \mathbb{E}[\mathbf{Q}(u)]\mathbb{E}\big[\mathbf{P}_q(u,t)\big]du, \tag{17}$$

and:

$$\mathbb{E}\big[\mathbf{P}_q(s,t)\big] = \mathbf{I} + \int_s^t \mathbb{E}\big[\mathbf{P}_q(u,t)\big]\mathbb{E}[\mathbf{Q}(u)]du. \tag{18}$$

A detailed solution of (17) and (18) could be found in [36], paragraph 8.9, where apparently $\mathbb{E}[\mathbf{Q}(t)]$ is a function of $\{Z(t)\}_{t\geq0}$ and $\{\mathcal{R}_I(t)\}_{t\geq0}$ due to the selection of $\mathbf{R}(t)$ by the compromise non-homogeneous Markov process. However, we are not interested in a closed analytic formula $\mathbb{E}\big[\mathbf{P}_q(s,t)\big]$, and it is sufficient that we know that it exists and that it is unique.

In what follows, I will use a probabilistic argument in order to find $\mathbb{E}[\mathbf{N}(t)]$, which will also be the solution of the differential Equation (13). The initial number of memberships $T(0) = \mathbf{N}(0)\mathbf{1}^{\top}$ at time t will be distributed to the various states with probabilities $\mathbb{E}\big[\mathbf{P}_q(0,t)\big]$, which are the probabilities of transitions of the expected embedded non-homogeneous Markov process generated by the intensity matrix $\mathbb{E}[\mathbf{Q}(t)]$. Thus, the expected distribution across the states of the initial memberships will be:

$$\mathbf{N}(0)\mathbb{E}\big[\mathbf{P}_q(0,t)\big]. \tag{19}$$

Now, let the time interval be $[x, x + \delta x)$, then the new memberships entering in that time interval are $D(x)\delta x$, and their expected values in the various states at the end of the interval are given by $\mathbf{p}_0(x, x + \delta x)D(x)\delta x$. After time $t - x$, the expected number of new memberships will be distributed to the various states of the population and their expected values will be $\mathbf{p}_0(x, x + \delta x)D(x)\delta x \mathbb{E}\big[\mathbf{P}_q(x,t)\big]$; therefore, integrating x from 0 to t, we get:

$$\mathbb{E}[\mathbf{N}(t)] = \mathbf{N}(0)\mathbb{E}\big[\mathbf{P}_q(0,t)\big] + \int_0^t \mathbf{p}_0(x)D(x)\mathbb{E}\big[\mathbf{P}_q(x,t)\big]dx, \tag{20}$$

4. The Asymptotic Behavior of the S-NHMSC

It is evident from previous studies, for example [1,4,35,37–40], that the central problems in the theory of NHMS and S-NHMS in discrete time, which will be studied in the present for S-NHMSC, are basically of two natures. The first classical problem is that of finding the conditions under which the asymptotic behavior of the expected population structure $\mathbb{E}[\mathbf{N}(t)]$ as $t \to \infty$ exists and finding its limit in closed analytic form as a function of the limits of the basic parameters of the system. The second classical problem is finding which expected relative population structures are possible limiting ones, provided that we control the limiting vector of input probabilities in the population.

In what follows, I will use as a norm of matrix $\mathbf{A} \in M_{k \times k}(\mathbb{R})$ the following:

$$\|\mathbf{A}\| = \sup_i \sum_i |a_{ij}|.$$

I will start by refreshing concepts and borrowing some important results from the theory of non-homogeneous Markov processes, starting with the following definitions for non-homogeneous Markov processes with countable state spaces.

Definition 1. *A Markov process* $\{X_t\}_{t=0}^{\infty}$ *is weakly ergodic if for every* $s \geq 0$, $\lim_{t \to \infty} \delta(\mathbf{P}(s,t)) = 0$.

In the case of weak ergodicity the probability of the occurrence of any of the states at time t tends to be independent from the initial probability distribution, but is in general dependent on t.

Definition 2. *A Markov process* $\{X_t\}_{t=0}^{\infty}$ *is ergodic if for every* $s \geq 0$, *there exists a vector* $\Pi = (\pi_1, \pi_2, ...)$ *such that:*

$$\lim_{t \to \infty} |p_{ij}(s,t) - \pi_j| = 0 \text{ for every } i, j \in \mathbb{S}.$$

Definition 3. *A Markov process* $\{X_t\}_{t=0}^{\infty}$ *is strongly ergodic if there exists a row-constant matrix* Π *such that, for all* $s \geq 0$:

$$\lim_{t \to \infty} \|\mathbf{P}(s,t) - \Pi\| = 0. \tag{21}$$

Remark 1. *When the state space* $\mathbb{S}$ *is finite, then the concepts of ergodic and strongly ergodic coincide.*

As the reader by now may have recognized, the generator of a non-homogeneous Markov process is the sequence of intensity matrices $\{\mathbf{Q}(t)\}_{t=0}^{\infty}$. This is so in the sense that the transition probability matrix $\mathbf{P}$ could be seen as the generator of a homogeneous Markov chain, and the sequence of transition probability matrices $\{\mathbf{P}(t)\}_{t=1}^{\infty}$ as the generator of a non-homogeneous Markov chain. Hence, our goal will now be to find conditions for strong ergodicity for a non-homogeneous Markov process based on the convergence of the sequence of intensity matrices $\{\mathbf{Q}(t)\}_{t=1}^{\infty}$.

I will now borrow a basic theorem concerning strong ergodicity for a non-homogeneous Markov chain based on its sequence of intensity matrices.

Theorem 1 ([14,15]). *Let a complete probability space be* $(\Omega, \mathcal{F}, \mathbb{P})$ *and a non-homogeneous Markov process* $\{X_t\}_{t=0}^{\infty}$ *with sequence of intensity matrices* $\{\mathbf{Q}(t)\}_{t=0}^{\infty}$, *which is such that* $\sup_{t \geq 0} \|\mathbf{Q}(t)\| \leq c$. *Let also a homogeneous Markov process be* $\{\hat{X}_t\}_{t=0}^{\infty}$ *with intensity matrix* $\mathbf{Q}$, *such that* $\|\mathbf{Q}\| \leq c$, *and which is strongly ergodic. If* $\lim_{t \to \infty} \|\mathbf{Q}(t) - \mathbf{Q}\| = 0$, *then if* Π *is the stable stochastic matrix, the limit of* $\{\hat{X}_t\}_{t=0}^{\infty}$, *then* $\{X_t\}_{t=0}^{\infty}$ *is also strongly ergodic with limit* Π.

Remark 2. *At this point, let us refresh the fact that for finite homogeneous, discrete, or continuous Markov chains, the concept of ergodicity, strong ergodicity, and weak ergodicity coincide. For an infinite chain, the notions of ergodicity and strong ergodicity are separated.*

I will present an important result from [16]. Let $\mathbf{Q}$ be the intensity matrix of a homogeneous Markov process $\{X_t\}_{t=0}^{\infty}$ and $\sup_{i \in \mathbb{S}}\{|q_{ii}|\} < c < \infty$ and $b > c$, define:

$$\hat{\mathbf{P}} = \mathbf{I} + \frac{\mathbf{Q}}{b},$$

then $\hat{\mathbf{P}}$ generates a discrete Markov chain $\{\hat{X}_t\}_{t=0}^{\infty}$.

Theorem 2 ([16]). *Let a complete probability space be* $(\Omega, \mathcal{F}, \mathbb{P})$ *and a finite homogeneous Markov process* $\{X_t\}_{t=0}^{\infty}$, *then it is ergodic if and only if the Markov chain* $\{\hat{X}_t\}_{t=0}^{\infty}$ *generated by* $\hat{\mathbf{P}} = \mathbf{I} + \mathbf{Q}/b$ *is ergodic.*

I will now prove the following basic theorem:

Theorem 3. *Let a complete probability space be* $(\Omega, \mathcal{F}, \mathbb{P})$ *and a finite S-NHMSC, as defined in Section* 2. *Assume that the following conditions hold:*

$$(1)\ \lim_{t\to\infty}\left\|\mathbf{R}_j(t) - \mathbf{R}_j\right\| = 0,\ (2)\ \lim_{t\to\infty}\left\|\mathbf{r}_{k+1}^\top(t) - \mathbf{r}_{k+1}^\top\right\| = 0,$$

$$(3)\ \lim_{t\to\infty}\left\|\mathbf{p}_0(t) - \mathbf{p}_0\right\|,\ (4)\ \lim_{t\to\infty}\left\|\mathbf{Z}(t) - \mathbf{Z}\right\| = 0,\ with$$

$$\sup_{t\geq0}\left\|\mathbf{Z}(t)\right\| < \infty,\ \sup_{i\in I}\{|z_{ii}|\} < z < \infty\ and\ let\ \mathbf{P}_Z = \mathbf{I} + \frac{\mathbf{Z}}{c_1}$$

with $c_1 > z$, $\mathbf{P}_Z$ *an irreducible, aperiodic matrix*

$$(5)\ \sup_{i\in S}\{r_{j,ii}\} < a < \infty,\ \sup\{r_{i,k+1}\} < b < \infty$$

where $r_{j,ii}$ *the* (i, j) *element of* $\mathbf{R}_j$,

then, as $t \to \infty$ $\mathbb{E}[\mathbf{Q}(t)]$ *converges in norm to the intensity matrix:*

$$\mathbb{E}[\mathbf{Q}] = \sum_{j=1}^{v} \pi_{z_j}\mathbf{R}_j + \mathbf{r}_{k+1}^\top\mathbf{p}_0,$$

where $\mathbf{\Pi}_Z = \left(\pi_{z_1}, \pi_{z_2}, ..., \pi_{z_k}\right)$ *is the left eigenvector of the eigenvalue* 1 *of the matrix* $\mathbf{P}_Z$.

Proof. From condition (4), since $\mathbf{P}_Z$ is an irreducible, aperiodic stochastic matrix, then there exists a stable stochastic matrix $\mathbf{\Pi}_Z$ with common row $\mathbf{\Pi}_Z = \left(\pi_{z_1}, \pi_{z_2}, ..., \pi_{z_k}\right)$, which is the left eigenvector of the eigenvalue 1 of the matrix $\mathbf{P}_Z$, that is:

$$\lim_{t\to\infty}\left\|\mathbf{P}_Z^t - \mathbf{\Pi}_Z\right\| = 0. \tag{22}$$

Furthermore, from condition (4), we have that the intensity matrices $\{\mathbf{Z}(t)\}_{t\geq0}$ converge to the intensity matrix $\mathbf{Z}$, and from (22), we know that it generates an ergodic Markov process. Therefore, $\{\mathbf{Z}(t)\}_{t\geq0}$, due to Theorem 2, generates an ergodic non-homogeneous Markov process, and we have that:

$$\lim_{t\to\infty}\left\|\mathbf{C}(s, t) - \mathbf{\Pi}_Z\right\| = 0. \tag{23}$$

We have that:

$$\mathbb{E}[\mathbf{R}(t)] = \sum_{i=1}^{v}\sum_{j=1}^{v} c_{ij}(0, t)c_i(0)\mathbf{R}_j(t). \tag{24}$$

Now, consider:

$$\left\|\mathbb{E}[\mathbf{Q}(t)] - \sum_{j=1}^{v}\pi_{z_j}\mathbf{R}_j - \mathbf{r}_{k+1}^\top\mathbf{p}_0\right\| \leq$$

$$\left\|\mathbb{E}[\mathbf{R}(t)] - \sum_{j=1}^{v}\pi_{z_j}\mathbf{R}_j\right\| + \left\|\mathbf{r}_{k+1}^\top(t)\mathbf{p}_0(t) - \mathbf{r}_{k+1}^\top\mathbf{p}_0\right\| \leq$$

$$\left\|\sum_{i=1}^{v}\sum_{j=1}^{v} c_{ij}(0, t)c_i(0)\mathbf{R}_j(t) - \sum_{i=1}^{v}\sum_{j=1}^{v}\pi_{z_j}c_i(0)\mathbf{R}_j\right\| +$$

$$\left\|\mathbf{r}_{k+1}^\top(t) - \mathbf{r}_{k+1}^\top\right\| + \left\|\mathbf{r}_{k+1}^\top\right\|\left\|\mathbf{p}_0(t) - \mathbf{p}_0\right\| \leq$$

$$\sum_{i=1}^{v}\sum_{j=1}^{v}\left\|c_{ij}(0, t)\mathbf{R}_j(t) - \pi_{z_j}\mathbf{R}_j\right\|c_i(0) +$$

$$\left\|\mathbf{r}_{k+1}^\top(t) - \mathbf{r}_{k+1}^\top\right\| + \left\|\mathbf{r}_{k+1}^\top\right\|\left\|\mathbf{p}_0(t) - \mathbf{p}_0\right\| \leq$$

$$\sum_{i=1}^{v}\sum_{j=1}^{v}\left[\left\|c_{ij}(0,t)-\pi_{z_j}\right\|\left\|\mathbf{R}_j(t)\right\|+\left\|\mathbf{R}_j(t)-\mathbf{R}_j\right\|\right]$$

$$+\left\|\mathbf{r}_{k+1}^{\top}(t)-\mathbf{r}_{k+1}^{\top}\right\|+\left\|\mathbf{r}_{k+1}^{\top}\right\|\left\|\mathbf{p}_0(t)-\mathbf{p}_0\right\|. \tag{25}$$

We have that:

$$\left\|\mathbf{R}_j(t)\right\|\leq\left\|\mathbf{R}_j(t)-\mathbf{R}_j\right\|+\left\|\mathbf{R}_j\right\|,$$

and since $\left|r_{j,ii}\right|=\sum_{l\neq i}\left|r_{j,il}\right|+\left|r_{j,i\,k+1}\right|$, and by condition (5) we have $\sup_{i\in\mathbb{S}}\{r_{j,ii}\}<a<\infty$, we could easily prove that:

$$\left\|\mathbf{R}_j\right\|\leq 2\left\{\sum_{l\neq i}\left|r_{j,il}\right|\right\}+\sup_{i\in\mathbb{S}}\left|r_{j,i\,k+1}\right|<b<\infty. \tag{26}$$

By condition (1), one can choose t_0^* such that for $t>t^*$, $\left\|\mathbf{R}_j(t)-\mathbf{R}_j\right\|<1$. Let $M^*=\sup_{0\leq t\leq t^*}\{\left\|\mathbf{R}_j(t)-\mathbf{R}_j\right\|\}$, denoted by $M=M^*+1+b$. Then:

$$\left\|\mathbf{R}_j(t)\right\|<M<\infty. \tag{27}$$

From (25), (27), and the conditions of the Theorem, we get that for $t>t_0$:

$$\left\|\mathbb{E}[\mathbf{Q}(t)]-\sum_{j=1}^{v}\pi_{z_j}\mathbf{R}_j+\mathbf{r}_{k+1}^{\top}\mathbf{p}_0\right\|\leq\epsilon.$$

Furthermore, it is not difficult using the conditions of the Theorem to see that:

$$\mathbb{E}[\mathbf{Q}]\mathbf{1}^{\top}=\sum_{j=1}^{v}\pi_{z_j}\mathbf{R}_j\mathbf{1}^{\top}+\mathbf{r}_{k+1}^{\top}\mathbf{p}_0\mathbf{1}^{\top}=-\sum_{j=1}^{v}\pi_{z_j}\mathbf{r}_{k+1}^{\top}+\mathbf{r}_{k+1}^{\top}=0.$$

$\square$

In analogy with the discrete case for an S-NHMS, we provide the following definition:

Definition 4. *We say that an S-NHMSC has an asymptotically attainable expected relative population structure $\mathbb{E}[\mathbf{q}(\infty)]$ under asymptotic input control, if there exists a $\mathbf{p}_0=\lim_{t\to\infty}\mathbf{p}_0(t)$ such that $\lim_{t\to\infty}\mathbb{E}[\mathbf{q}(t)]=\mathbb{E}[\mathbf{q}(\infty)]$. We denote by $\mathcal{A}^{\infty}$ the set of asymptotically expected relative population structures under asymptotic input control of the S-NHMSC.*

We now provide the following basic theorem concerning the asymptotic behavior of the S-NHMSC.

Theorem 4. *Let a complete probability space be $(\Omega,\mathcal{F},\mathbb{P})$ and a finite S-NHMSC as defined in Section 2. Assume that the conditions (1) $\to$ (5) of Theorem 2 hold and, in addition, that the following conditions are true (6):*

$$\lim_{t\to\infty}T(t)=T,$$

where $T(t)$ is a non-dicreasing continuous function. (7) The matrix:

$$\mathbf{P}_q=\mathbf{I}+\frac{\mathbb{E}[\mathbf{Q}]}{c_2},$$

with $c_2 > \sup_{i \in \mathbb{S}} \{|\mathbb{E}(q_{ii})|\}$ is an irreducible, aperiodic stochastic matrix. Then, (i) as $t \to \infty$, $\mathbb{E}[\mathbf{q}(t)]$ converges to $\Pi_q = \left(\pi_{q1}, \pi_{q2}, ..., \pi_{qk} \right)$, which is the left eigenvector of the eigenvalue 1 of the matrix $\mathbf{P}_q$. (ii) The set $\mathcal{A}^\infty$ is the convex hull of the points:

$$\mu_i \left[\mathbf{e}_i \left(\sum_{j=1}^{v} \pi_{z_j} \mathbf{R}_j \right)^{-1} \right], \text{ where } \mu_i = \mathbf{e}_i \left(\sum_{j=1}^{v} \pi_{z_j} \mathbf{R}_j \right)^{-1} \mathbf{1}^\top.$$

Proof. Since Π_z is the left eigenvector of the eigenvalue 1 of the irreducible, aperiodic matrix $\mathbf{P}_Z$, we have that $0 \leq \pi_{z_j} \leq 1$ for $j = 1, 2, ..., v$. Furthermore, condition (5) of Theorem 3 is also valid for the present; hence, $\sup_{i \in \mathbb{S}} \{r_{j,ii}\} < \infty$ and $\sup\{r_{i,k+1}\} < b < \infty$. Consequently, from the expression of $\mathbb{E}[\mathbf{Q}]$ in Theorem 3 we get that:

$$c_2 > \sup_{i \in \mathbb{S}} \{\mathbb{E}[q_{ii}]\} < \infty. \tag{28}$$

Now, since $\mathbf{P}_q$ is an irreducible, aperiodic stochastic matrix, we have that:

$$\lim_{t \to \infty} \left\| \mathbf{P}_q^t - \Pi_q \right\| = 0,$$

where Π_q is a stable stochastic matrix with row $\Pi_q = \left(\pi_{q1}, \pi_{q2}, ..., \pi_{qk} \right)$, which is the left eigenvector of the eigenvalue 1 of the matrix $\mathbf{P}_q$. From (28), Theorems 1 and 2, we have that if we denote with $\mathbb{E}\left[\mathbf{P}_q(s,t)\right]$ the probability transition matrix of the non-homogeneous Markov process defined by the intensities $\{\mathbb{E}[\mathbf{Q}(t)]\}_{t \geq 0}$, then:

$$\lim_{t \to \infty} \left\| \mathbb{E}\left[\mathbf{P}_q(s,t)\right] - \Pi_q \right\| = 0 \text{ for every } s \in \mathbb{N}. \tag{29}$$

Therefore, as $t \to \infty$ is the first part of the right hand side of Equation (20):

$$\lim_{t \to \infty} \mathbf{N}(0)\mathbb{E}\left[\mathbf{P}_q(s,t)\right] = \mathbf{N}(0)\Pi_q = T(0)\Pi_q. \tag{30}$$

Now, consider:

$$\mathbf{U}(t) = \left\| \int_0^t \mathbf{p}_0(x)D(x)\mathbb{E}\left[\mathbf{P}_q(x,t)\right]dx - \int_0^t \mathbf{p}_0 D(x)\Pi_q dx \right\|$$

$$\leq \int_0^t \left\| \mathbf{p}_0(x)\mathbb{E}\left[\mathbf{P}_q(x,t)\right] - \mathbf{p}_0 \Pi_q \right\| D(x)dx$$

$$\leq \int_0^t \left\| \mathbf{p}_0(x) \right\| \left\| \mathbb{E}\left[\mathbf{P}_q(x,t)\right] - \Pi_q \right\| D(x)dx + \int_0^t \left\| \mathbf{p}_0(x) - \mathbf{p}_0 \right\| D(x)dx$$

$$= \int_0^t \left\| \mathbb{E}\left[\mathbf{P}_q(x,t)\right] - \Pi_q \right\| D(x)dx + \int_0^t \left\| \mathbf{p}_0(x) - \mathbf{p}_0 \right\| D(x)dx$$

$$= \mathbf{A}(t) + \mathbf{B}(t)$$

From (29), we have that there exists a $t_0 > 0$ such that for $t - x > t_0$:

$$\left\| \mathbb{E}\left[\mathbf{P}_q(x,t)\right] - \Pi_q \right\| < \epsilon.$$

Thus:

$$\mathbf{A}(t) \leq \int_0^{t-t_0} \left\| \mathbb{E}\left[\mathbf{P}_q(x,t)\right] - \Pi_q \right\| D(x)dx +$$

$$\int_{t-t_0}^t \left\| \mathbb{E}\left[\mathbf{P}_q(x,t)\right] - \Pi_q \right\| D(x)dx$$

$$\leq \epsilon\{T(t - t_0) - T(0)\} + 2\{T(t) - T(t - t_0)\}. \tag{31}$$

Now, from condition (3), we have that there exists a $t_1 > 0$ such that for $t > t_1$, $\|\mathbf{p}_0(t) - \mathbf{p}_0\| < \epsilon$. Thus:

$$\mathbf{B}(t) = \int_0^t \|\mathbf{p}_0(x) - \mathbf{p}_0\| D(x) dx \le$$

$$\int_0^{t_1} \|\mathbf{p}_0(x) - \mathbf{p}_0\| D(x) dx + \int_{t_1}^t \|\mathbf{p}_0(x) - \mathbf{p}_0\| D(x) dx$$

$$\le 2(T(t_1) - T(0)) + \epsilon(T(t) - T(t_1)).$$

Hence, we have that $\lim_{t\to\infty} \mathbf{A}(t) = 0$ and $\lim_{t\to\infty} \mathbf{B}(t) < \infty$. Thus, $\mathbf{U}(t)$ is an increasing function bounded from above and $\lim_{t\to\infty} \mathbf{U}(t) = 0$. Therefore, from (31), we have that:

$$\lim_{t\to\infty} \int_0^t \mathbf{p}_0(x) D(x) \mathbb{E}\big[\mathbf{P}_q(x,t)\big] dx = \int_0^t \mathbf{p}_0 D(x) \mathbf{\Pi}_q dx$$

$$= \mathbf{p}_0 \mathbf{\Pi}_q [T - T(0)] = \mathbf{\Pi}_q [T - T(0)]. \tag{32}$$

Hence, from (20), (30) and (32), we get that:

$$\lim_{t\to\infty} \mathbb{E}[\mathbf{N}(t)] = \mathbb{E}[\mathbf{N}(t)] = T\mathbf{\Pi}_q. \tag{33}$$

Since $\|\mathbb{E}[\mathbf{Q}]\|$ is finitely bounded and defines an ergodic Markov process, it is known that:

$$\mathbf{\Pi}_q \mathbb{E}[\mathbf{Q}] = 0. \tag{34}$$

From Theorem 3 and Equation (34), we get that

$$\mathbf{\Pi}_q \sum_{j=1}^{v} \pi_{z_j} \mathbf{R}_j = -\mathbf{\Pi}_q \mathbf{r}_{k+1} \mathbf{p}_0 = \mathbf{\Pi}_q \mathbf{R}_j \mathbf{1}^\top \mathbf{p}_0. \tag{35}$$

The matrix $\sum_{j=1}^{v} \pi_{z_j} \mathbf{R}_j$, due to condition (7), is irreducible and aperiodic and is part of the intensity matrix $\mathbb{E}[\mathbf{Q}]$. Hence, ([41]) $\left(\sum_{j=1}^{v} \pi_{z_j} \mathbf{R}_j\right)^{-1}$ exists and is nonnegative. Therefore:

$$\mathbf{\Pi}_q = \mathbf{\Pi}_q \mathbf{R}_j \mathbf{1}^\top \mathbf{p}_0 \left(\sum_{j=1}^{v} \pi_{z_j} \mathbf{R}_j\right)^{-1}, \tag{36}$$

and:

$$\mathbf{\Pi}_q = \mathbf{\Pi}_q \mathbf{R}_j \mathbf{1}^\top \sum_{i=1}^{k} p_{0i} \mathbf{e}_i \left(\sum_{j=1}^{v} \pi_{z_j} \mathbf{R}_j\right)^{-1}. \tag{37}$$

Multiplying both sides of (37) by $\mathbf{1}^\top$, we obtain:

$$1 = \mathbf{\Pi}_q \mathbf{R}_j \mathbf{1}^\top \sum_{i=1}^{k} p_{0i} \mathbf{e}_i \left(\sum_{j=1}^{v} \pi_{z_j} \mathbf{R}_j\right)^{-1} \mathbf{1}^\top. \tag{38}$$

Let:

$$\mu_i = \mathbf{e}_i \left(\sum_{j=1}^{v} \pi_{z_j} \mathbf{R}_j\right)^{-1} \mathbf{1}^\top. \tag{39}$$

Then:

$$1 = \mathbf{\Pi}_q \mathbf{R}_j \mathbf{1}^\top \sum_{i=1}^{k} p_{0i} \mu_i. \tag{40}$$

Therefore, from (33) and the above, we get that:

$$\lim_{t \to \infty} \mathbb{E}[\mathbf{q}(t)] = \mathbf{\Pi}_q = \sum_{i=1}^{k} \frac{p_{0i}\mu_i}{\sum_{j=1}^{k} p_{0j}\mu_j} \mu_i^{-1} \left[\mathbf{e}_i \left(\sum_{j=1}^{v} \pi_{z_j} \mathbf{R}_j \right)^{-1} \right]. \tag{41}$$

Hence, $\mathbb{E}[\mathbf{q}(\infty)]$ is a convex combination of the vertices:

$$\mu_i^{-1} \left[\mathbf{e}_i \left(\sum_{j=1}^{v} \pi_{z_j} \mathbf{R}_j \right)^{-1} \right].$$

$\square$

It is well known that for a homogeneous Markov process, with intensity matrix $\mathbf{Q}$ and transition matrix $\mathbf{P}(t)$, which is strongly ergodic, the rate of convergence with which $\mathbf{P}(t)$ converges to a stable stochastic matrix is exponential. Logically, this fact creates the intuition, that possibly for a non-homogeneous Markov process with sequence of intensity matrices $\mathbf{Q}(t)$, the rate at which the transition probability matrices converge to a stable stochastic matrix is also exponential. The answer to this is negative, since we need one more condition for this to be true, and that is $\lim_{t \to \infty} \|\mathbf{Q}(t) - \mathbf{Q}\| = 0$ with an exponential rate of convergence. This result is stated formally in the following theorem, the proof of which could be found in [14].

Theorem 5. *Let a complete probability space be $(\Omega, \mathcal{F}, \mathbb{P})$ and a non-homogeneous Markov process $\{X_t\}_{t=0}^{\infty}$ with sequence of intensity matrices $\{\mathbf{Q}(t)\}_{t=0}^{\infty}$, which is strongly ergodic. Let also a homogeneous Markov process be $\{\hat{X}_t\}_{t=0}^{\infty}$ with intensity matrix $\mathbf{Q}$, which is strongly ergodic. Let $g : \mathbb{R}^+ \to \mathbb{R}^+$ be a monotonically increasing function. If $\lim_{t \to \infty} g(2t)\|\mathbf{Q}(t) - \mathbf{Q}\| = 0$ then:*

$$\lim_{t \to \infty} \sup_{s \geq 0} \{\min(\exp(\lambda t)), g(t)\|\mathbf{P}(s,t) - \mathbf{\Pi}\|\} = 0,$$

where $0 < \lambda < \beta/2$ and $\beta > 0$ is the constant parameter of the exponential rate of convergence at which $\{\hat{X}_t\}_{t=0}^{\infty}$ converges.

An important question which logically arises is: what is the rate of convergence to asymptotically attainable structures in an S-NHMSC? In fact, I am interested in finding conditions under which the rate is exponential, because then, the practical value of the asymptotic result is greater (see [42,43]). Furthermore, as in [20], the problem of construction of sharp bounds for the rate of convergence of characteristics of Markov chains to their limiting vectors is very important. That is, all too often, it is easier to calculate the limit characteristics of a process than to find the exact distribution of state probabilities. Therefore, it is very important to have a possibility to use the limit characteristics as asymptotic approximations for the exact distribution. The following Theorem answers the question of the rate of convergence of the expected structure of an S-MHMSC.

Theorem 6. *Let a complete probability space be $(\Omega, \mathcal{F}, \mathbb{P})$ and a finite S-NHMSC as defined in Section 2. Furthermore, let the conditions (1) $\to$ (7) of Theorem 4 hold and in addition assume that the convergences in conditions (1) $\to$ (4) and (6) are exponentially fast. Then, the convergence of $\mathbb{E}[\mathbf{N}(t)]$ as $t \to \infty$ is exponentially fast.*

Proof. Since $\lim_{t \to \infty} \|\mathbf{Z}(t) - \mathbf{Z}\| = 0$ is exponentially fast and in addition $\mathbf{Z}$ is strongly ergodic, then in Theorem 5 there are constants c_3 and $\lambda_1 > 0$ such that:

$$\|\mathbf{C}(s, s+t) - \mathbf{\Pi}_z\| \leq c_3 e^{-\lambda_1 t} \text{ for every } s, t > 0. \tag{42}$$

Since the convergences in conditions (1)–(3) are exponentially fast, we have that:

$$\exists\, c_0 > 0,\, a_0 > 0 \text{ such that } \|\mathbf{p}_0(t) - \mathbf{p}_0\| \le c_0 e^{-a_0 t} \text{ for every } t. \tag{43}$$

$$\exists\, c_1 > 0,\, a_1 > 0 \text{ such that } \left\|\mathbf{r}_{k+1}^\top(t) - \mathbf{r}_{k+1}^\top\right\| \le c_1 e^{-a_1 t} \text{ for every } t. \tag{44}$$

$$\exists\, c_2 > 0,\, a_2 > 0 \text{ such that } \|\mathbf{R}_j(t) - \mathbf{R}_j\| \le c_2 e^{-a_2 t} \text{ for every } t. \tag{45}$$

From (25), (42) → (45), we arrive at:

$$\|\mathbb{E}[\mathbf{Q}(t)] - \mathbb{E}[\mathbf{Q}]\| \le c e^{-at} \text{ with } c > 0, a > 0. \tag{46}$$

Now, from (46), condition (7), of Theorems 4 and 5 we get that:

$$\left\|\mathbb{E}\left[\mathbf{P}_q(s, s+t) - \mathbf{\Pi}_q\right]\right\| \le c_q e^{-\lambda_2 t}, c_q, \lambda_2, t > 0 \text{ for every } t. \tag{47}$$

We now have the following:

$$\left\|\mathbb{E}[\mathbf{N}(t)] - T\mathbf{\Pi}_q\right\| = \left\|\mathbf{N}(0)\mathbb{E}\left[\mathbf{P}_q(0, t)\right]\right. \tag{48}$$

$$+ \int_0^t \mathbf{p}_0(x) D(x) \mathbb{E}\left[\mathbf{P}_q(x, t)\right] dx - T\mathbf{\Pi}_q \Big\| \le$$

$$\|\mathbf{N}(0)\| \left\|\mathbb{E}\left[\mathbf{P}_q(0, t)\right] - \mathbf{\Pi}_q\right\| +$$

$$\left\|\int_0^t \mathbf{p}_0(x) D(x) \mathbb{E}\left[\mathbf{P}_q(x, t)\right] dx - (T - T(0))\mathbf{\Pi}_q\right\|$$

$$\le \|\mathbf{N}(0)\| \left\|\mathbb{E}\left[\mathbf{P}_q(0, t)\right] - \mathbf{\Pi}_q\right\| + \left\|(T(t) - T)\mathbf{\Pi}_q\right\| +$$

$$\left\|\int_0^t \mathbf{p}_0(x) D(x) \mathbb{E}\left[\mathbf{P}_q(x, t)\right] dx - (T(t) - T(0))\mathbf{\Pi}_q\right\| \le$$

$$\|\mathbf{N}(0)\| \left\|\mathbb{E}\left[\mathbf{P}_q(0, t)\right] - \mathbf{\Pi}_q\right\| + |(T(t) - T)| +$$

$$\int_0^t \left\|\mathbb{E}\left[\mathbf{P}_q(x, t)\right] - \mathbf{\Pi}_q\right\| D(x) dx + \int_0^t \|\mathbf{p}_0(x) - \mathbf{p}_0\| D(x) dx \tag{49}$$

From (47), condition (3), we obtain the fact that the convergence as $t \to \infty$ of $T(t)$ is exponentially fast, and based on (49), we arrive at the following relation:

$$\left\|\mathbb{E}[\mathbf{N}(t)] - T\mathbf{\Pi}_q\right\| \le c e^{-\lambda t} \text{ with } c, \lambda, t > 0 \text{ and for every } t > 0,$$

which proves the Theorem. □

5. An Illustrative Example from Manpower Planning

In the present section, the previous results are illustrated through an example from manpower planning. Interesting examples of such systems can be found in [44]. Suppose that intensities were estimated from the historical records of a firm with three grades, and they found that three were repeatedly exercised; thus, the pool $\mathcal{R}_1(t)$ has the elements:

$$\mathbf{R}_1(t) = \begin{pmatrix} -4 - 2e^{-3t} & 3 + e^{-3t} & 0 \\ 0 & -5 - 3e^{-t} & 3 + 2e^t \\ 0 & 0 & -7 - e^{-5t} \end{pmatrix},$$

$$\mathbf{R}_2(t) = \begin{pmatrix} -5 - 10e^{-3t} & 4 + 9e^{-3t} & 0 \\ 0 & -6 - 9e^{-t} & 4 + 7e^t \\ 0 & 0 & -7 - e^{-5t} \end{pmatrix},$$

$$\mathbf{R}_1(t) = \begin{pmatrix} -3 - 4e^{-3t} & 2 + 3e^{-3t} & 0 \\ 0 & -7 - 3e^{-t} & 5 + e^{t} \\ 0 & 0 & -7 - e^{-5t} \end{pmatrix}.$$

Let also:

$$\mathbf{r}_{k+1}(t) = \left[1 + e^{-3t}, 2 + e^{-7}, 7 + e^{-5t} \right], \quad \mathbf{p}_0(t) = (0.2 \quad 0.3 \quad 0.5).$$

In addition, let us utilize the well-known maximum likelihood estimates for transition intensities ([44]); the matrix of the transition intensities of the compromise non-homogeneous Markov process $\{\mathbf{Z}(t)\}_{t \geq 0}$, under the assumption that they are time independent, was found to be:

$$\hat{\mathbf{Z}} = \begin{pmatrix} -5 & 3 & 2 \\ 4 & -9 & 5 \\ 2 & 5 & -8 \end{pmatrix}.$$

Applying Theorem 3 to the above data, we have that conditions (1)–(3) are satisfied with:

$$\mathbf{R}_1 = \begin{pmatrix} -4 & 3 & 0 \\ 0 & -5 & 3 \\ 0 & 0 & 7 \end{pmatrix}, \mathbf{R}_2 = \begin{pmatrix} -5 & 4 & 0 \\ 0 & -6 & 4 \\ 0 & 0 & 7 \end{pmatrix},$$

$$\mathbf{R}_3 = \begin{pmatrix} -3 & 2 & 0 \\ 0 & -7 & 5 \\ 0 & 0 & -7 \end{pmatrix} \text{ and } \mathbf{r}_{k+1} = (1 \quad 2 \quad 7).$$

Obviously, $\sup_{t \geq 0} \|\mathbf{Z}(t)\| < \infty$, and with $c_1 = 10$, we get:

$$\mathbf{P}_Z = \begin{pmatrix} 0.5 & 0.3 & 0.2 \\ 0.4 & 0.1 & 0.5 \\ 0.3 & 0.5 & 0.2 \end{pmatrix},$$

which is obviously an irreducible regular stochastic matrix, and thus, condition (4) and condition (5) of Theorem 3 are satisfied. Now, the asymptotic expected intensity matrix is found to be:

$$\mathbb{E}[\mathbf{Q}] = \begin{pmatrix} -3.8 & 3.3 & 0.5 \\ 0.4 & -5.3 & 4.9 \\ 1.4 & 2.1 & -3.5 \end{pmatrix},$$

which, apparently, is a matrix of transition intensities.

Theorems 4 and 5 are straightforwardly applicable with the above data. The present example could be used as a guide for applying the theoretical results in many areas of potential applications, such as for example in [44–48].

6. Conclusions

The concept of a non-homogeneous Markov system in a stochastic environment and in continuous time was introduced. It was found under which conditions, using basic parameters, the limiting population structure and the relating relative population structure exist, and they were evaluated in elegant closed analytic forms. The set of all possible relative population structures was characterized under all possible input probability vectors. Finally, an illustrative example from manpower planning was presented, which could be used as a guide for applications in other areas.

Funding: This research received no external funding.

Institutional Review Board Statement: Not applicable.

Informed Consent Statement: Not applicable.

Data Availability Statement: Not applicable.

Conflicts of Interest: The author declares no conflict of interest.

References

1. Tsantas, N.; Vassiliou, P.-C.G. The non-homogeneous Markov system in a stochastic environment. *J. Appl. Probab.* **1993**, *30*, 285–301. [CrossRef]
2. Choen, J.E. Ergodicity of age structure in populations with Markovian vital rates, I: Countable states. *J. Am. Stat. Assoc.* **1976**, *71*, 335–339.
3. Choen, J.E. Ergodicity of age structure in populations with Markovian vital rates, II: General states. *Adv. Appl. Probab.* **1977**, *9*, 18–37.
4. Vassiliou, P.-C.G. Asymptotic behavior of Markov systems. *J. Appl. Probab.* **1982**, *19*, 815–817. [CrossRef]
5. Bartholomew, D.J. *Stochastic Models for Social Processes*, 3rd ed.; Wiley: New York, NY, USA, 1982.
6. Gani, J. Formulae for projecting enrolments and degrees awarded in universities. *J. R. Stat. Soc.* **1963**, *126*, 400–409. [CrossRef]
7. Colnisk, J. Interactive Markov chains. *J. Math. Sociol.* **1976**, *6*, 163–168.
8. Young, A.; Vassiliou, P.-C.G. A non-linear model for the promotion of staff. *J. R. Stat. Soc.* **1974**, *137*, 584–595. [CrossRef]
9. Vassiliou, P.-C.G. A Markov chain model for wastage in manpower systems. *Oper. Res. Q.* **1976**, *27*, 57–70. [CrossRef]
10. Vassiliou, P.-C.G. A high order non-linear Markovian model for promotion in Manpower systems. *J. R. Stat. Soc.* **1978**, *141*, 86–94. [CrossRef]
11. Iosifescu, M. *Finite Markov Processes and Applications*; John Wiley: New York, NY, USA, 1980.
12. Goodman, G.S. An intrinsic time for non-stationary Markov chains. *Z. Wahrscheinlichkeitsth.* **1970**, *16*, 165–180. [CrossRef]
13. Scott, M.; Arnold, B.C.; Isaacson, D.L. Strong ergodicity for continuous time non-homogeneous Markov chains. *J. Appl. Probab.* **1982**, *19*, 692–694. [CrossRef]
14. Johnson, J.T. Ergodic Properties of Non-Homogeneous continuous Markov Chains. Ph.D. Thesis, Iowa State University, Ames, IA, USA, 1984.
15. Johnson, J.T.; Isaacson, D. Conditions for strong ergodicity using intensity matrices. *J. Appl. Probab.* **1988**, *25*, 34–42. [CrossRef]
16. Yong, P.L. Some results related to q-bounded Markov processes. *Nanta Math.* **1976**, *8*, 34–41.
17. Zeifman, A.I. Quasi-ergodicity for non-homogeneous continuous time Markov chains. *J. Appl. Probab.* **1989**, *26*, 643–648. [CrossRef]
18. Zeifman, A.I.; Isaacson, D.L. On strong ergodicity for non-homogeneous continuous-time Markov chains. *Stoch. Process. Their Appl.* **1994**, *50*, 263–273. [CrossRef]
19. Zeifman, A.I.; Korolev, V.Y. Two sided bounds on the rate of convergence for continuous-time finite inhomogeneous Markov chains. *Stat. Probab. Lett.* **2015**, *137*, 84–90. [CrossRef]
20. Zeifman, A.I.; Korolev, V.Y.; Satin, Y.A.; Kiseleva, K.M. Lower bounds for the rate of convergence for continuous-time inhomogeneous Markov chains with a finite state space. *Stat. Probab. Lett.* **2018**, *103*, 30–36. [CrossRef]
21. Mitrophanov, A.Y. Stability and exponential convergence of continuous time Markov chains. *J. Appl. Probab.* **2003**, *40*, 970–979. [CrossRef]
22. Bartholomew, D.J. *Stochastic Models for Social Processes*, 2nd ed.; Wiley: New York, NY, USA, 1973.
23. McClean, S.I. A continuous time population model with Poisson recruitment. *J. Appl. Probab.* **1976**, *13*, 348–354. [CrossRef]
24. McClean, S.I. Continuous time stochastic models of a multigrade population. *J. Appl. Probab.* **1978**, *15*, 26–37. [CrossRef]
25. Gerontidis, I. On certain aspects of non-homogeneous Markov systems in continuous time. *J. Appl. Probab.* **1990**, *27*, 530–544. [CrossRef]
26. McClean, S.I.; Montgomery, E.; Ugwuowo, F. Non-homogeneous continuous-time Markov and semi-Markov manpower models. *Appl. Stoch. Models Data Anal.* **1998**, *13*, 191–198. [CrossRef]
27. Tsaklidis, G. The evolution of the attainable structures of a continuous time homogeneous Markov system with fixed size. *J. Appl. Probab.* **1996**, *33*, 34–47. [CrossRef]
28. Kipouridis, I.; Tsaklidis, G. The size order of the state vector of continuous-time homogeneous Markov system with fixed size. *J. Appl. Probab.* **2001**, *38*, 635–646. [CrossRef]
29. Vasiliadis, G.; Tsaklidis, G. On the distribution of the state sizes of closed continuous time homogeneous Markov systems. *Method. Comput. Appl. Probab.* **2009**, *11*, 561–582. [CrossRef]
30. Vasiliadis, G. On the distributions of the state sizes of the continuous time homogeneous Markov system with finite state capacities. *Methodol. Comput. Appl. Probab.* **2012**, *14*, 863–882. [CrossRef]
31. Vasiliadis, G. Transient analysis of the M/M/k/N/N queue using acontinuous time homogeneous Markov system with finite state size capacity. *Commun. Stat. Theory Methods* **2014**, *43*, 1548–1562. [CrossRef]
32. Dimitriou, V.A.; Georgiou, A.C. Introduction, analysis and asymptotic behavior of a multi-level manpower planning model in a continuous time setting under potential department contraction. *Commun. Stat. Theory Methods* **2020**, *50*, 1173–1199. [CrossRef]
33. Esquivel, M.L.; Krasil, N.P.; Guerreiro, G.R. Open Markov type population models: From discrete to continuous time. *Mathematics* **2021**, *9*, 1496. [CrossRef]
34. Bartholomew, D.J. Maintaining a grade or age structure in a stochastic environment. *Adv. Appl. Prob.* **1977**, *11*, 603–615. [CrossRef]
35. Vassiliou, P.-C.G. The evolution of the theory of non-homogeneous Markov systems. *Appl. Stoch. Models Data Anal.* **1997**, *13*, 159–176. [CrossRef]

36. Iosifescu, M. *Finite Markov Processes and Applications*; Dover Publications: New York, NY, USA, 2007.
37. Georgiou, A.C.; Vassiliou, P.-C.G. Periodicity of asymptotically attainable structures in Non-homogeneous Markov systems. *Linear Algebra Its Appl.* **1992**, *176*, 137–174. [CrossRef]
38. Tsantas, N.; Georgiou, A.C. Periodicity of equilibrium structures in a time dependent Markov model under stochastic environment. *Appl. Stoch. Models Data Anal.* **1994**, *10*, 269–277. [CrossRef]
39. Tsantas, N. Ergodic behavior of a Markov chain model in a stochastic environment. *Math. Methods Oper. Res.* **2001**, *54*, 101–117. [CrossRef]
40. Vassiliou, P.-C.G.; Tsantas, N. Stochastic Control in Non-Homogeneous Markov Systems. *Int. J. Comput. Math.* **1984**, *16*, 139–155. [CrossRef]
41. Darroch, J.N.; Seneta, E. On quasi-stationary distribution in absorbing continuous-time finite Markov chains. *J. Appl. Probab.* **1988**, *25*, 34–42.
42. Vassiliou, P.-C.G.; Tsaklidis, G. The rate of convergence of the vector of variances and covariances in non-homogeneous Markov set systems. *J. Appl. Probab.* **1989**, *27*, 776–783. [CrossRef]
43. Vassiliou, P.-C.G. On the periodicity of non-homogeneous Markov chains and systems. *Linear Algebra Its Appl.* **2015**, *471*, 654–684. [CrossRef]
44. Bartholomew, D.J.; Forbes, A.; McClean, S.I. *Statistical Techniques in Manpower Planning*; Wiley: Chichester, UK, 1991.
45. McClean, S.I. Using Markov models to characterize and predict process target compliance. *Mathematics* **2021**, *9*, 1187. [CrossRef]
46. McClean, S.I.; Gillespie, J.; Garg, L.; Barton, M.; Scotney, B.; Fullerton, K. Using phase-type models to cost stroke patient care across health, social and community services. *Eur. J. Oper. Res.* **2014**, *236*, 190–199. [CrossRef]
47. Patoucheas, P.D.; Stamou, G. Non-homogeneous Markovian models in ecological modeling a study of zoobenthos in Thermaikos Gulf, Greece. *Ecol. Modell.* **1993**, *66*, 197–215. [CrossRef]
48. Gao, K.; Yan, X.; Peng, R.; Xing, L. Economic design of a linear consecutive connected system considering cost and signal loss. *IEEE Trans. Syst. Man Cybern. Syst.* **2021**, *51*, 5116–5128. [CrossRef]

Article

Introducing a Novel Method for Smart Expansive Systems' Operation Risk Synthesis

Nikolay Zhigirev [1], Alexander Bochkov [2,*], Nataliya Kuzmina [3] and Alexandra Ridley [4]

[1] KALABI IT, 107045 Moscow, Russia; nnzhigirev@mail.ru
[2] JSC NIIAS, 109029 Moscow, Russia
[3] Department "Organization of Transportation on Air Transport", Moscow State Technical University of Civil Aviation, 125493 Moscow, Russia; n.kuzmina@mstuca.aero
[4] Department of Computational Informatics and Programming, Institute of Information Technology and Applied Mathematics, Moscow Aviation Institute (National Research University), 125993 Moscow, Russia; sunridl@gmail.com
* Correspondence: a.bochkov@vniias.ru

Abstract: In different areas of human activity, the need to choose optimal (rational) options for actions from the proposed alternatives inevitably arises. In the case of retrospective statistical data, risk analysis is a convenient tool for solving the problem of choice. However, when planning the growth and development of complex systems, a new approach to decision-making is needed. This article discusses the concept of risk synthesis when comparing alternative options for the development of a special class of complex systems, called smart expansive systems, by the authors. "Smart" in this case implies a system capable of ensuring a balance between its growth and development, considering possible external and internal risks and limitations. Smart expansive systems are considered in a quasi-linear approximation and in stationary conditions of problem-solving. In general, when the alternative to comparison is not the object itself, but some scalar way of determining risks, the task of selecting the objects most at risk is reduced to assessing the weights of factors affecting the integral risk. As a result, there is a complex task of analyzing the risks of objects, solved through the amount by which the integral risk can be minimized. Risks are considered as anti-potentials of the system development, being retarders of the reproduction rate of the system. The authors give a brief description of a smart expansive system and propose approaches to modeling the type of functional dependence of the integral risk of functioning of such a system on many risks, measured, as a rule, in synthetic scales of pairwise comparisons. The solution to the problem of reducing the dimension of influencing factors (private risks) using the vector compression method (in group and inter-scale formulations) is described. This article presents an original method for processing matrices of incomplete pairwise comparisons with indistinctly specified information, based on the idea of constructing reference-consistent solutions. Examples are provided of how the vector compression method can be applied to solve practical problems.

Keywords: decision-making; expert assessments; pairwise comparisons; risk; smart expansive system; uncertainty; vector compression method

Citation: Zhigirev, N.; Bochkov, A.; Kuzmina, N.; Ridley, A. Introducing a Novel Method for Smart Expansive Systems' Operation Risk Synthesis. *Mathematics* **2022**, *10*, 427. https://doi.org/10.3390/math10030427

Academic Editor: Victor Korolev

Received: 14 December 2021
Accepted: 26 January 2022
Published: 28 January 2022

Publisher's Note: MDPI stays neutral with regard to jurisdictional claims in published maps and institutional affiliations.

1. Introduction

Global processes of transformation have begun in recent years. The actuality of solving the tasks of effective management of structurally complex sociotechnical systems has increased significantly. The concept of the value of objects (assets) included in such systems "presses up" the value of assessments and the significance of decision-making becomes determining, but the definition of this "value" remains more an art than a scientifically based methodology. Intuitively, with limited resources (of all kinds), it is necessary to aim to use these resources in the most rational way. However, to develop a rational solution, it is necessary to learn how to assess the results of the targeted activities of the system, how to apply them to the tasks set and learn the costs associated with each solution. To make a comparison, you need to learn how to measure certain quantitative features that characterize the functionality of an individual object and the entire complex sociotechnical system. You also need a "tool" that allows you to assess the result. Since the task lies in choosing the best of the compared options for the growth and development of the system, first, it is necessary to learn how to measure the quality of decision-making. The quality of any solution becomes apparent only in the process of its implementation (in the process of target operation of a managed object or system). Therefore, the main objective is the assessment of the quality of the solution by the effectiveness of its application. Thus, to reasonably choose the preferred solution, it is necessary to measure the effectiveness of the target function of the managed object or system versus its compared variants [1] in conditions of existing uncertainty and risk.

The comparison of options and decision-making directly depends on the competence of decision-makers (DM), that is, on their ability to comprehensively assess the risks associated with the functioning and development of the system. To ensure a well-founded choice of DM, we propose using decision-formers (DF), which as a rule, are analytical tools based on mathematical methods. These can be found in Kantorovich's "simplex method" [2] on through to modern machine-learning methods, neural networks [3–5], methods of reference vectors [6], genetic algorithms [7], etc.

There are several classes of decision-making tasks:

- deterministic, which are characterized by a one-valued connection between the decision-making and its outcome, aimed at building the "progress" function and determining the stable parameters at which the optimum is achieved;
- stochastic, in which each decision made can lead to one of many outcomes occurring with a certain probability. Usually, it uses simulation programming methods [8], game theory [9] or other methods of adaptive stochastic management [10] to choose the optimal strategy in view of averaged, statistical characteristics of random factors;
- in conditions of uncertainty, when the criterion of optimality depends, in addition to the strategies of the operating party and fixed risk factors, and also on uncertain factors of a non-stochastic nature, an interval mathematics [11] approach, or approximations in the form of fuzzy (blurred) sets [12,13], are used in decision-making.

The latter case involves processing the views of independent experts [14,15]. Despite the wide application of expert systems in practice, the fairness of using certain methods of analysis remains incomprehensible for many DM, especially when the results are contrary to "common sense" (in their understanding) [16]. So, developers have to formulate and uphold certain principles, without which the automation of methods adopted in expert systems becomes unacceptable.

Often, expert assessment procedures are based on the method of processing matrices of pairwise comparisons of various alternatives, known as the Saaty algorithm (or hierarchy analysis method) [17]. It is quite widely used despite criticism [18–20] and the lack of a one-valued solution to several research issues.

Firstly, with large dimensions for the pairwise comparison matrix, the number of comparisons for each expert increases to $N \times (N-1)/2$, where N is the number of alternatives considered. Problems arise with the "poor-quality" filling of the comparison matrix by experts and the "insufficient" quality scale used in the method.

Secondly, not all experts can compare all proposed alternatives in pairs, and so some pairwise comparison matrices will remain unassessed (NA). Partially, this issue was solved by Saaty with the development of the method of analysis of hierarchies and the method of analytical networks, but the latter contains several strong assumptions that impose restrictions on its application [21,22].

Thirdly, as a rule, there is no "reference" alternative; the remaining assessments are obtained by converting $A_{i,j} = A_{i,1} \times A_{j,1}^{-1}$, which is used, for example, in combinatorial methods for restoring missing data [19].

Fourthly, when summarizing the opinions of experts and moving to a common matrix of pairwise comparisons, values with significant variation appear in the same cells, which necessitates working with the assessments set in the interval scale [23].

Finally, in the case where the alternative to comparison is not the object itself, but instead a scalar method is determining risks, then the task of selecting objects is reduced to assessing the weights of factors affecting the integral risk. As a result, a complex problem of analyzing the risks of objects arises, which can be solved by minimizing the integral risk [24].

2. Smart Expansive System

Complex systems theory (synergy) uses nonlinear modeling and fractal analysis for forecasting. In the last decade, such innovative areas as theoretical history and mathematical modeling of history, based on a synergistic, holistic description of society as a non-linear developing system, have been actively developing (V. Glushkov, B. Onyky, N. Zhigirev, S. Kurdyumov, D. Chernavsky, V. Belavin, S. Malkov, A. Malkov, V. Korotaev, D. Khalturina, P. Turchin, V. Budanov).

Modern complex sociotechnical systems are characterized by distribution in space, a large variety of objects included in their composition and interaction of various types of objects, a heterogeneous structure of transport and technological chains, unique conditions for influencing individual objects and the system as a combination of risks of various natures. If the stability of the functioning of such complex systems means the implementation of their development plan with permissible deviations in terms of volume and timing of tasks, then their management is reduced to minimizing unscheduled losses in emergency situations and taking measures to anticipate them, that is, for the analysis, assessment and management of associated risks.

The concept of management of such systems strives to achieve an optimal balance between the cost of the object, associated risks and performance indicators, based on which economic goals are formed and the use of the object is ensured in such a way that it creates added value. In general, optimal profit-oriented management strikes a balance in the reallocation of available resources (material, human and information) between "productive activities" and "maintaining development potential".

The above is described most closely by models of interaction between a developing object and its environment, such as in the model of self-perfecting developing systems of V. Glushkov. He introduced a new class of dynamic models based on nonlinear integral-differential equations with a history [25]. He also developed approaches to modeling the so-called "self-perfecting systems" and proved theorems on the existence and uniqueness of solutions, describing their systems of equations [26].

However, it should be noted that the name "developing", applied to the class of systems in question, is not quite correct and contains some ambiguous interpretations. The growth of the system may not be accompanied by its development (for example, improving the science of creation and design, instructions for the manufacture or use of the product) and vice versa (for example, expectations of a quick practical return on basic science). Usually, growth and development are combined, there is a smooth or uneven change in the proportions between them and some "equilibrium" state with the external environment occurs (or does not occur).

In parallel with the work of V. Glushkov, works on Scientific and Technical Progress (STP) have also begun in specific industries. Examples are the studies of B. Onyky and V. Reznichenko [27,28], who laid the foundations for the theory of potential systems, and the early works of N. Zhigirev [29]. Based on biophysical and economic models, they proposed a practical new version including integral-differential equations that describe the process of producing, introducing and forgetting knowledge in production cycles due to the transition to other scientific and technological foundations. They demonstrated the cyclical nature of capacity-building and the need to develop complex systems (health, education, industrial safety systems, ecology and other infrastructure projects by generation).

Our understanding and development of the aforementioned models compel us to introduce into consideration a class of so-called "smart expansive systems," consisting of three subsystems (Figure 1).

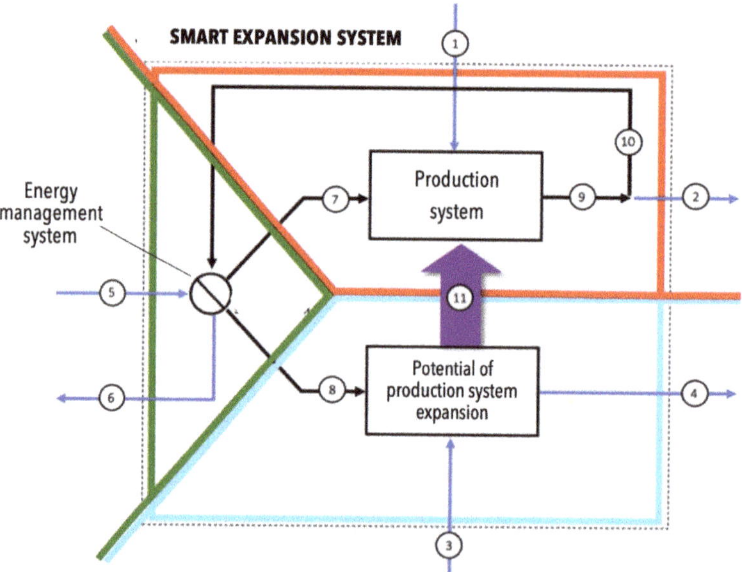

Figure 1. Scheme of smart expansive system functioning.

The smart expansive system (SES) is an open system that may be growing or developing, or simultaneously both growing and developing (when, for example, objects of different "generations" are included in it). Sometimes the growth of the system is accompanied by "flattening" (flatlining at the same level of development) and degradation in its development. The SES is open since it needs to effectively allocate the necessary resources from the external environment, possibly "cleaning" them before use (including the human resource), and remove biowaste.

The production subsystem is assessed by the reproduction rate multiplier at a conditional minimum development potential. Below this minimum (critical mass of potential), the growth and development of the SES are impossible in principle. The potential catalytic function describing this multiplier in the limit is an asymptotic curve with saturation (like a logistic curve), although potential inhibition is also possible since the production subsystem occupies the space of a common SES. This behavior is similar to the flow system of the "Brusselator" (an intensively working conveyor) when initial substrates flow out of it, which do not have time to react with the catalyst, despite their sufficient amount [30].

The production subsystem serves to measure the success of expansion, determined, for example, by the volume of useful products produced by the system.

The subsystem of the expansion potential of the production is intended for catalytic management of the produced forms and resources (sometimes measured by money with a dual structure—the cost of renewing matter and the cost of maintaining information in the broad sense of the word).

The energy management subsystem is actually a two-circuit resource management system (financial and temporary) between production and contribution to infrastructure projects.

In Figure 1, the externally directed energy flow (5) is distributed by the control subsystem to the production subsystem (7) and goes to the expansion potential subsystem for the production of knowledge and improvement of technologies, which are "recipes" for the preparation of products (the so-called flow to the development of "infrastructure").

From the external environment, the expansion potential subsystem also receives additional information on new knowledge, inventions and technologies (3) and has a catalytic effect on the production subsystem (11). The production subsystem, in turn, receives from the external environment a flow of "purified" semi-finished products (1) for further expansion. In the process of expansion, there is inevitably a partial forgetting of information due to various causes, including the physical death of the carriers of the original thought forms (4), which causes a weakening of the expansion potential of the entire system.

There is an outflow (2) of products from the production subsystem to the external environment, including unused semi-finished products, waste from the assembly of products, etc. Purified from (2), the flow of energy from the results of labor to the production subsystem and the results of sales of products on the market supports the functioning of the regulatory subsystem. In the latter, over time, it is possible to disperse energies (6) not yet distributed among subsystems that can cause, under certain conditions, a collapse of the management system.

Leaving beyond the scope of this article a detailed description of external and internal interactions, let us dwell a little more on the features of a deterministic and stochastic approach to modeling smart expansive systems.

2.1. Deterministic Model of Smart Expansive System

For the deterministic case, the SES is described by a two-parameter model for time (1) and for the proportions of the energy distribution (3). The first equation describing the system is in a sense autonomous:

$$\frac{dX(t)}{dt} = \left(g \times \frac{\varphi(\beta)}{1+\beta} - a \right) \times X(t) - b \times X^2(t), \tag{1}$$

where $X(t)$ is the volume of the "production subsystem" measured by the number of products; $a \times X(t)$—the additive of linear part—maintains the production technology and requires linear costs (in economics, for example, these are depreciation costs); $g \times \frac{\varphi(\beta)}{1+\beta} \times X(t)$ is a linear production function of a useful subsystem with the parameter β; $s = \frac{1}{1+\beta}$ ($0 \le \beta \le \infty$) is the proportion of energy distribution from newly created forms s ($s \in [0,1]$; $(1-s) = \frac{\beta}{1+\beta}$ ($0 \le \beta \le \infty$); g is the coefficient of the scale of production losses, where as a rule, $0 \le g \le 1$ is performed; $\varphi(\beta)$ beforehand is the set amplifier of the production of forms due to reading "the correct information" (e.g., instructions for assembly), with information as a catalytic function; $b \times X^2(t)$ is a quadratic term that considers the limited "semi-finished products" and the competition of finished "products" in the surrounding world.

The function $\varphi(\beta)$ takes the form of a logistic curve (Figure 2), which is generally unnecessary if the requirements of positive constrained monotonicity are met. This function can have breaks of the first kind. The final form of the function $\varphi(\beta)$ with the argument β is also determined by the degree of necessary detail for the calculation.

Segments on the abscissa axis $[\beta_1; \beta_2]$, $[\beta_2; \beta_L]$ and $[\beta_R; \beta_5]$ if $X_k(\beta) < 0$ are the area of degradation of the smart expansive system. Accordingly, the segments $[0; \beta_1]$, $[\beta_L; \beta^*]$ and $[\beta^*; \beta_R]$ if $X_k(\beta) > 0$ are the area of its growth (development). Moreover, the expansion of the system begins only from β_2 (on the segment $[\beta_2; \beta^*]$ $X_k(\beta)$ grows, and on the rest, it only decreases). Above the limit value $\beta = \beta_5$, it makes no sense to look for a solution, although already at the point β_R, the system begins to degrade actively.

The type of logistic curve is selected so that in the segment $[0; \beta_1]$, the efficiency of the productive subsystem is extremely low (this is the field of low-skilled labor and individual potentially breakthrough ideas in science). The segment $[\beta_L; \beta_R]$ corresponds to mass production using the available knowledge and skills. The optimal $X_k(\beta^*)$ is inside $[\beta_L; \beta_R]$, while at $[\beta_L; \beta^*]$, science is not sufficiently developed and highly-demanded, and at $[\beta^*; \beta_R]$, science is "too much" and the results of scientific research simply do not have time to be introduced and mastered in the production subsystem.

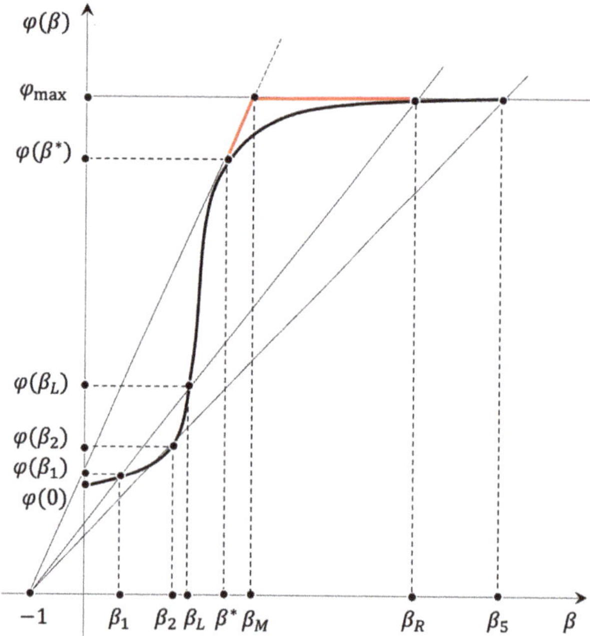

Figure 2. Dependence on "potential for development" from expired costs.

The point $\beta = 0$ in Figure 2 corresponds to a situation where all resources are spent exclusively on the growth of the production subsystem. The potential of such a system is low due to permanent losses that can be avoided if there is the potential to anticipate and manage emerging risks.

The section $\beta = (0, \beta_1)$ shows that if the funds assigned to study and counteract threats and risks are small, then the return on such research and activities is less than the resources assigned to them. Information collection for low-level investigation of internal and external threats does not allow for an adequate assessment that improves the quality of decision-making in most cases.

On the segment $\beta = (\beta_1, \beta^*)$, the contribution to the development potential begins to give a positive return; however, the so-called "self-repayment" level of the costs of developing the "potential" of the system $\varphi(\beta_L)$ will be achieved only at the point β_L. Therefore, it is advisable to consider this point as a point of "critical" position. The reduction of the potential $\varphi(\beta)$ to the level $\varphi(\beta_L)$ threatens a situation where "due to the circumstances," the "survival strategy" will be economically suitable—that is, taking the strategy of completely eliminating the cost of solving the tasks of prediction and anticipation of threats and risks, and ensuring reproduction only by increasing low-efficiency capacities in the production subsystem $\beta \to 0$.

The optimum is reached at point β^*. This point has a certain sense to it. If resources for the development of potential are given "excessively" $(\beta \succ \beta^*)$, then the funds $(\beta - \beta^*)$ are incorrectly removed from the current reproduction and a situation arises where disproportionate efforts are spent on studying and counteracting many risks that the developing system may never face. The optimum does not depend on the values g, a, b. There may be cases where a is so large that even with an optimal solution, the system does not develop, but degrades. The condition of non-degeneracy of the decision is the presence of positive ordinates in X_K (Figure 3).

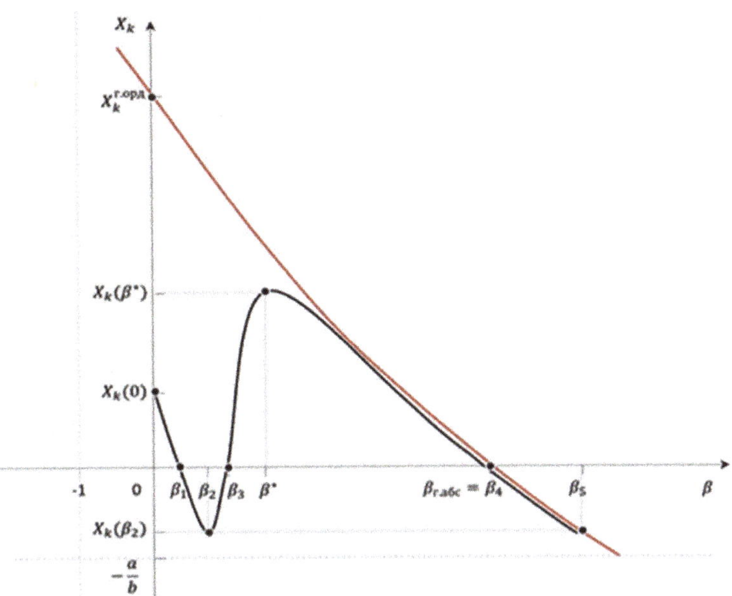

Figure 3. Dependency graph $X_k(\beta)$.

Generally, for the segment $[0, \beta_L]$, only the overregulation and excessive formalization of management of a system can do much harm, and for the segment $[\beta^*, \infty]$, there is a situation where the costs of searching for an existing solution are so great that it is preferable to receive it anew.

In Figure 2, with β along the abscissas axis and $\varphi(\beta)$ along the ordinate axis, we draw half-lines from the point $T(-1, 0)$ crossing the graph's $\varphi(\beta)$. The position of the half-line is set at $(z = \alpha(\beta + 1))$ with a tangent α to the abscissa axis β and crosses over the ordinate axis at point $(0, \alpha)$.

The position of the upper touch of the half-lines corresponds to the coordinate $(\beta^*, \varphi(\beta^*))$, for which the following is performed

$$\alpha^* = \frac{\varphi(\beta^*)}{(1 + \beta^*)} \tag{2}$$

With a fixed $\beta(s)$ and $\varphi(\beta)$, an initial $X(0) = X_0$ (next to zero deformation of the space of forms) and an asymptotically final volume $X(\infty) = X_K$, solutions can be represented in the form of a logistic curve.

If $a1 = \dfrac{\left(g \times \frac{\varphi(\beta)}{1+\beta} - a\right)}{b}$ and $X_K = \frac{a1}{b}$, then $X(t) = \dfrac{X_K \times X_0}{(X_K - X_0) \times \exp(-a1 \times t) + X_0}$.

Let us find out how the parameter β influences final productivity, which means a stable equilibrium state of X_K.

Considering the difference scheme, we define the final growth graph ΔX with a step Δt: $X(t + \Delta t) = X(t) + \Delta X(t) \times \Delta t$, where $\Delta X(t) = g \times \frac{\varphi(\beta)}{1+\beta} - a$.

The final graph $X_k(\beta)$, taking into account the compression of the graph $\varphi(\beta)$ by g and the displacement by the value a vertically downward, followed by division by the value b, is shown in Figure 3.

The second equation describes the dynamics of the laws of functioning of the system, by which $X(t)$ is created

$$\frac{dY(t)}{dt} = f \times (Y_0 - Y(t)) + h \times \frac{\beta \times \varphi(\beta)}{1 + \beta} \times X(t). \tag{3}$$

Here, Y_0 is the initial state. The regulator f describes the loss of knowledge, but $\left(h \times \frac{\beta \times \varphi(\beta)}{1+\beta}\right) \times X(t)$ corresponds to the complication of laws for new unknown management functions ("emergence") that are not available in subsystems.

We define the maximum Y_K :

$$Y_K = Y_0 + \frac{hg}{bf} \times \frac{\beta \times \varphi(\beta)}{1+\beta} \times \left(\frac{\varphi(\beta)}{1+\beta} - A\right), \tag{4}$$

where the parameter $A = \frac{a}{g}$.

$$Y_K = Y_0 + \beta \times \frac{hg}{bf} \times \left[+\frac{A}{2} + \frac{\varphi(\beta)}{1+\beta} - \frac{A}{2}\right] \times \left[-\frac{A}{2} + \frac{\varphi(\beta)}{1+\beta} - \frac{A}{2}\right], \tag{5}$$

$$Y_K = Y_0 + \beta \times \frac{ha^2}{4 \times gbf} \times \left[+1 + \left(\frac{2}{A}\frac{\varphi(\beta)}{(1+\beta)} - 1\right)\right] \times \left[-1 + \left(\frac{2}{A}\frac{\varphi(\beta)}{(1+\beta)} - 1\right)\right]. \tag{6}$$

Making a replacement

$$C = \left[\left(\frac{2}{A} \times \frac{\varphi(\beta)}{(1+\beta)} - 1\right)^2 - 1\right] \times \beta, \tag{7}$$

we get the final solution of the following form

$$Y_K = Y_0 + \frac{ha^2}{4 \times gbf} \times C, \tag{8}$$

where parameter C is searched for as the maximum value from the above range.

Maximum C is reached either at the edges of the segment or in one of the local minima. So, in Figure 2, on the asymptote, $\beta \in [\beta_M, \beta_R]$, $\varphi(\beta) = \varphi_{max}$ is performed.

This allows us to obtain a four-parametric model $\{\beta^*, \varphi(\beta^*), \beta_R, \varphi_{max}\}$ of the initial quasi-linear growth.

$$Z = \frac{2 \times g \times \varphi_{max}}{a}; \, s = \frac{1}{(1+\beta)}; \tag{9}$$

$$C = \left[(Z \times s - 1)^2 - 1\right] \times \frac{(1-s)}{s} = -\left(Z \times s - \left(\frac{Z}{2} + 1\right)\right)^2 + \left[\frac{Z}{2} - 1\right]^2 \tag{10}$$

$$x_{min} = 0; \, s_R = \frac{1}{(1+\beta_R)} = \frac{a}{g \times \varphi_{max}}; C_R = 0 \tag{11}$$

$$s_{opt} = \frac{1}{2} + \frac{1}{Z}; \, C_{opt} = \left[\frac{Z}{2} - 1\right]^2 = \left[\frac{g \times \varphi_{max}}{a} - 1\right]^2 > 0; \, s^* = \frac{1}{(1+\beta^*)} \tag{12}$$

$$s_M = \frac{1}{(1+\beta_M)} = \frac{\varphi(\beta^*) \times s^*}{\varphi_{max}} \tag{13}$$

Here, it is possible to give an explicit formula for C_M

$$C_M = \frac{4}{(\varphi_{max} \times s_R)^2}\{1 - s_M\} \times \{s_M - s_R\} > 0, \tag{14}$$

and it is possible to limit this to a simplified form

$$C_M = -\left(Z \times s_M - \left(\frac{Z}{2} + 1\right)\right)^2 + \left[\frac{Z}{2} - 1\right]^2 < C_{opt} \tag{15}$$

In conclusion, we confirm that s_{opt} gets into the segment $[s_R, s_M]$, i.e.,

$$s_R \leq s_{opt} = \frac{1}{2} + \frac{1}{Z} \leq s_M. \tag{16}$$

The solutions are either s_M or s_{opt} depending on

$$\left(\frac{1}{2} + \frac{1}{Z} - s_R\right)\left(s_M - \frac{1}{2} - \frac{1}{Z}\right) \geq 0, \tag{17}$$

$$\left(1 - \frac{a}{g \times \varphi_{max}}\right)\left(\frac{\varphi(\beta^*) \times s^*}{\varphi_{max}} - \frac{1}{2} - \frac{a}{2 \times g \times \varphi_{max}}\right) \geq 0. \tag{18}$$

The first bracket in (18) is the inability to create anything $(\beta > \beta_R)$—in this case, it acts as a criterion for the feasibility of expansion in general.

The second bracket, subject to

$$\frac{\varphi(\beta^*) \times s^* \times g - a}{b} = x_{max}, \tag{19}$$

$$\left(\frac{x_{max} \times b}{g \times \varphi_{max}} - \frac{1}{2} + \frac{a}{2 \times g \times \varphi_{max}}\right) \geq 0 \tag{20}$$

is a criterion for achieving the desired performance x_{max} in the production subsystem.

The positivity of their work guarantees

$$s_{opt} = \frac{1}{2} + \frac{a}{2 \times g \times \varphi_{max}}; \ C_{opt} = \left[\frac{g \times \varphi_{max}}{a} - 1\right]^2. \tag{21}$$

Otherwise, the optimum is on the left edge s_M, i.e.,

$$C_{opt} = C_M. \tag{22}$$

However, there are unique systems, the value of which depends solely on the capacity of the producer. When the potential is high and its assessment is not underestimated relative to "fair"—i.e., $g \gg 0$—the second bracket is carried out, even at $x_{max} = 1$ and with a lack of competition from other producers $b \cong 0$.

The presence of two optimal solutions β^* (by the number of products) and β^{opt} (by the "mind"), determined through the constant C_{opt} in (22), means that optimal will be $\beta^{opt} > \beta^*$ refusal of the gross product in favor of maximum use of the expansion potential, that is, production with the "optimal" margin of "possible use" of the manufactured products (multifunctionality).

Despite the counter nature of the described model, this gives the idea that threats and risks can be considered as "anti-potentials" of development (i.e., they are retardants of the reproduction rate of the entire system). To model a real system, it is necessary to analyze the "raw" process data and then synthesize them into a meaningful structure explaining the process under study.

2.2. Stochastic Linear Growth Models of Smart Expansive System

2.2.1. Model of System Growth Taking into Account the Effect of Random Perturbations of System Productivity on the Speed of Its Reproduction

The model takes into account that in the "quasi-linear section" of the expansion of the system, not only is the speed of expansion important but also the dispersion of the process. At the same time, the "volatility" of the process itself plays a greater role than the profitability of the "production subsystem".

Despite an increase in the "average" amount of product from each element of the system, each element is individually characterized by a limited time of effective operation. Moreover, the indicator of "population mortality" under natural restrictions on mathematical expectation is mainly influenced by the magnitude of the variance. Therefore, for example, in economics, where processes with mathematical expectation values of several percent are studied, the variance values themselves appear in the definitions of "risks".

Here, it is extremely important to note that to assess the values of mathematical expectation and variance, they are quantified at the starting point of time based on group assessments. It is further hypothesized that these assessments obtained for the group can be used to predict the trajectories of each element of the group separately. This is a very strong assumption since the model claims, firstly, that the obtained assessments will remain constant for the entire forecast time, and secondly, it is established that each element at any time behaves in the same way as some element at zero point in time. Such assumptions are true, generally speaking, only for ergodic processes. Yet, not all the processes described by the model in question are ergodic. In systems consisting of elements of more than one type, the need to consider such "risks" is greatly increased, and these "risks" themselves are much higher.

2.2.2. Model of Impact of Capital Fluctuation on System Growth

This model is not related to the properties of the system itself, but to the level of fluctuations in the parameters (influencing factors) of the external environment (fluctuations in the level of corruption, changes in tax legislation, etc.). The most likely values for the number of elements in such a model are always less than its average value. A certain value is introduced as the threshold of state criticality.

If the current value is lower than the critical value, the probability of ruin increases sharply. It is important to note that with the increase in time, critical values also increase.

Moreover, if the fluctuation amplitude of the distribution variance assessment is large and the mathematical expectation and initial value are sufficiently small, then the probability of degeneration of the system tends to one.

Thus, on average, external fluctuations accelerate the growth of the system, but the payment for such accelerated growth is an increased probability of its degeneration (a decrease in the mathematical expectation of its degeneration time), and since the expansion process is multifactorial, but the "history" of the behavior of such a system (as is the case, for example, in mass service systems), as a rule, is not, it is essential that rather than conducting an analysis based on statistics from past observation periods, instead, a synthesis of the risk of functioning of the "smart expansive system" is carried out.

3. Smart Expansive System's Risk Synthesis

Regarding risk, the concept of "synthesis" is currently hardly used in contrast to the concept of "analysis". However, it is necessary to understand that risk analysis is characteristic of systems in which risk realization events occur often enough to apply a well-developed apparatus of probability theory and mathematical statistics. This approach works in insurance, for example, in the theory of reliability, when we deal with the flow of insurance cases, accidents or breakdowns. Yet, when it comes to ensuring safety in an era where the main characteristic is constant unsteadiness and variability, it is possible to do this only through the synthesis of risks, developing automated advising systems that become complicated as they develop tips for professionals (DF) or replacing professionals with highly intelligent robotic systems. The risk from the concept of "analytics" in this case is becoming "synthetic".

As the analysis of integrated assessments of the state of complex objects and systems used in system studies shows, generalized risk criteria (indices) are widely used. There are additive (weighted average arithmetic) and multiplicative (weighted average geometric) forms of these:

- arithmetic (smoothing "emissions" of private risk indicators) $R_{ar} = \sum_{i=1}^{M}(\alpha_i \times r_i)$;
- geometric (enhancing negative "emissions" of private risk rating) $R_{ge} = \prod_{i=1}^{M} r_i^{\alpha_i}$;
- geometric anti-risk $1 - R_\varnothing = U_\varnothing = \prod_{i=1}^{M}(u_i)^{\alpha_i} = \prod_{i=1}^{M}(1 - r_i)^{\alpha_i}$.

Weight coefficients α_i of partial estimates r_i satisfy the condition

$$\sum_{i=1}^{N} \alpha_i = 1; \; \alpha_i > 0 \; (i = 1, \dots, M). \tag{23}$$

Real numbers r_i (private risks) take values from the interval between zero and one.

For smart expansive systems, the most acceptable form of risk representation is geometric anti-risk [31], which satisfies the main a priori requirements underlying the risk approach to the construction of a nonlinear integral assessment of $R_\varnothing$, namely:

1. smoothness—the continuous dependence of the integral R assessment and its derivatives on private assessments: $R(r_1, \dots, r_M)$;
2. limitation—limits of the interval of change of private r_i and integral R assessments:

$$0 < R(r_1, \dots, r_M) < 1 \, ; \; \text{if } 0 < r_1, r_2, \dots, r_M < 1. \tag{24}$$

3. equivalence—the same importance of private assessments r_i and r_j;
4. hierarchical single-level—aggregate only the private assessments of r_i that belong to the same level of the hierarchical structure;
5. neutrality—the integral assessment coincides with the private assessment when the other one takes the minimum value:

$$R(r_1,\, 0) = r_1; R(0, r_2) = r_2; R(0,\, 0) = 0; R(1,\, 1) = 1. \tag{25}$$

6. uniformity $R(r_1 = r, \dots, r_M = r) = r$.

The geometric anti-risk derives from the concept of "difficulties in achieving the goal" proposed by I. Russman [31], and is the "assessment from above" for the weighted average arithmetic and weighted average geometric risk.

Risk as a measure of the "difficulty in achieving the goal," and assesses the difficulty of obtaining the declared result d_k with the existing resource quality assessments (μ_k) and requirements for this quality (ε_k). The concept of difficulty in achieving a goal with a given quality and given requirements for the quality of a resource and the result follows from the considerations that it is more difficult to obtain a result of a certain quality when there is a low quality of the resource or a high requirement for its quality.

For general reasons, the difficulty of obtaining d_k result should have the following basic properties:

- when $\mu_k = \varepsilon_k$ should be maximum, i.e., equal to one (indeed, the difficulty of obtaining the result is maximum at the lowest permissible quality value);
- if $\mu_k = 1$ and $\mu_k >> \varepsilon_k$ should be minimal, that is, equal to zero (for the maximum possible value of quality, regardless of the requirements (for $\varepsilon_k < 1$), the complexity should be minimal);
- if $\mu_k > 0$ and $\varepsilon_k = 0$ should be minimal, that is, equal to zero (obviously, if there are no requirements for the quality of the resource components and μ_k is more than zero, then the difficulty of obtaining a result for this component should be minimal).

For these three conditions for $\varepsilon_k < \mu_k$, a function of type $d_k = \frac{\varepsilon_k(1-\mu_k)}{\mu_k(1-\varepsilon_k)}$ is allowed.

We also assume that $d_k = 0$ for $\mu_k = \varepsilon_k = 0$ and $d_k = 1$ for $\mu_k = \varepsilon_k = 1$.

The functioning of the reliable system is characterized by the preservation of its main characteristics within the established limits. The actions of such a system are aimed at minimizing deviations of its current state from some given ideal goal. In relation to the system, the goal can be considered as the desired state of its outcome, that is, not only the value of its objective function.

Let us briefly explain the essence of geometric anti-risk. We will consider the system in the process of achieving the goal, moving from its current state to some future result, the quantitative expression of which is A_{pl}. Let us suppose that the goal is achievable in time t_{pl}. We also assume that there is a minimum speed of movement $v_{\min}$ to the goal in time and a maximum speed v_{max}. It is most convenient to measure the quantitative expression of the result and the time required to achieve it in dimensionless values; to do this, we assume A_{pl} and t_{pl} equate to 1 or 100%. In Figure 4, the minimum and maximum speed trajectories of the system correspond to the OD and OB lines.

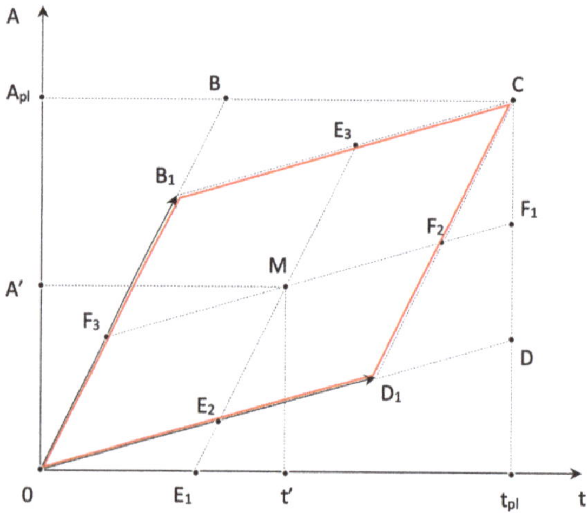

Figure 4. Geometric interpretation of system movement toward goal.

Polyline OD_1C is the boundary of the exclusion zone, and for any point M with coordinates (t', A'), to describe the position of the system on an arbitrary trajectory to the goal within the parallelogram OB_1CD_1, the distance $r(M)$ is taken as the risk of not reaching the goal:

$$r(M) = max\left\{ \ln \frac{1}{1-d_1}, \ \ln \frac{1}{1-d_2} \right\} \tag{26}$$

where $d_1 = \frac{\varepsilon_1(1-\mu_1)}{\mu_1(1-\varepsilon_1)}$, $d_2 = \frac{\varepsilon_2(1-\mu_2)}{\mu_2(1-\varepsilon_2)}$, $\varepsilon_1 = \frac{|E_1 E_2|}{|E_1 E_3|}$, $\mu_1 = \frac{|E_1 M|}{|E_1 E_3|}$, $\varepsilon_2 = \frac{|F_1 F_2|}{|F_1 F_3|}$, $\mu_2 = \frac{|F_1 M|}{|F_1 F_3|}$.

We emphasize that the geometric anti-risk satisfies, in addition, the so-called theorem "on the fragility of the good" in the theory of disasters, according to which, " ... for a system belonging to a special part of the stability boundary, with a small change in parameters, it is more likely to get into the area of instability than the area of stability". This is a manifestation of the general principle that everything good (for example, stability) is more fragile than everything bad [32]. Risk analysis uses a similar risk-limiting principle. Any system can be considered "good" if it meets a certain set of requirements, but must be considered "bad" if at least one of them is not fulfilled. At the same time, everything "good"—for example, the ecological security of territories—is more fragile, meaning it is easy to lose it and difficult to restore.

The continuous function $R(r_1,\ldots,r_i,\ldots,r_n)$ satisfying the above conditions has the following general form (27):

$$R(r_1,\ldots,r_i,\ldots,r_n) = 1 - \left\{ \prod_{i=1}^{n}(1-r_i) \right\} \times g(r_1,\ldots,r_i,\ldots,r_n), \tag{27}$$

If in a special case $g(r_1,\ldots,r_i,\ldots,r_n) \equiv 1$, then:

$$R(r_1,\ldots,r_i,\ldots,r_n) = 1 - \left\{ \prod_{i=1}^{n}(1-r_i) \right\}, \tag{28}$$

which gives an understated assessment of integral risk based on the calculation that the flow of abnormal situations for objects of the system is a mixture of ordinary events taken from homogeneous but different values of r_i $(i = 1, \ldots, n)$ samples.

Since for real systems, risks are usually dependent, we have

$$g(r_1, \ldots, r_i, \ldots, r_n) = 1 - \sum_{i=1}^{n-1} \sum_{j=i+1}^{n} C_{ij} \times [r_i]^{\alpha_{ij}} \times [r_j]^{\beta_{ij}}, \tag{29}$$

$$\sum_{i=1}^{n-1} \sum_{j=i+1}^{n} C_{ij} \leq 1, \; C_{ij} \geq 0, \; \alpha_{ij} > 0, \; \beta_{ij} > 0, \tag{30}$$

where C_{ij} are the risk connectivity coefficients of the i-th and j-th abnormal situations for the objects of the system and α_{ij} and β_{ij} are positive elasticity coefficients replacing the corresponding risks. These allow for taking into account the facts of "substitution" of risks, mainly since effective measures to reduce all risks cannot simultaneously be carried out due to the limited time and resources of DM.

The current values of private risks r_i $(i = 1, \ldots, n)$ included in the integral indicator (27) are values that vary over time at different speeds (for example, depending on the seasonal factor, the priorities of the solved technological problems in some systems of the fuel and energy complex change significantly).

Private risks r_i are built, as a rule, through the convolution of the corresponding resource indicators—factors of influence that have a natural or value expression. These factors are measured on certain synthetic scales (for example, in the previously mentioned multiplicative Saaty, the pairwise comparison scale), the mutual influences of which should also be studied since they are generally nonlinear and piecewise-continuous.

To obtain assessments of factors of influence, weighted scales must be built. To this end, the authors have developed the so-called "vector compression method" [33], which is discussed herein.

4. Vector Compression Method: Stationary Method of Incomplete Pairwise Comparison of Risk Factors

When moving from the preference scale to the linear logarithmic scale, the weights of objects $W(x_i)$ are converted into $v_i = \log(W(x_i))$, and the matrix of pairwise comparisons of the factors of influence of private risks $A = (a_{ij})$ is converted into an incomplete antisymmetric matrix $\overline{A} = (\log a_{ij})$ [21].

The indicator matrix monitors the status of the link network. It takes two values: $G_{i,j}[t] = 1$ when the link occurs, and $G_{i,j}[t] = 0$ when it is absent (NA, not available). This article deals only with the case of stationary matrices $G_{i,j}[t] = G_{i,j}$.

An analog of the matrix consistency condition in such a statement is the elements of the antisymmetric error matrix E_{ij}:

$$E_{ij} = \overline{a}_{ij} - v_i + v_j. \tag{31}$$

For correlation, matrices $E_{ij} = 0$ for all $i, j \in [1, N]$.

4.1. General Properties of Transformations of Antisymmetric Matrices

Let us introduce the basic designations:

$E^{i*}_{max}[t]$—maximum value of elements of matrix $E_{i,j}[t]$, for which indicator function $G_{i,j} = 1$ (maximum on line i);

$E^{i*}_{min}[t]$—minimum value of elements of matrix $E_{i,j}[t]$, for which indicator function $G_{i,j} = 1$ (minimum on line i);

$E^{*i}_{max}[t]$—maximum value of matrix elements $E_{j,i}[t]$, for which indicator function $G_{j,i} = 1$ (maximum by column i).

The essence of the transformation of the matrix $E[t+1] = S^i_x(E[t])$ with the parameter x consists of an element-by-element reduction by x of all elements of the i-th line and an increase by x of all elements of the i-th column of the antisymmetric matrix $E[t]$.

Transformations $E[t_1 + 1] = S^i_{x[t_1]}(E[t_1])$, $E[t_2 + 1] = S^i_{x[t_2]}(E[t_2])$, ..., differing only by the parameter $x[t]$ with the constant i, we will call the same type.

The vector compression process is illustrated in Figure 5, in which the abscissa axis is the parameter x and the ordinate axes are two deposited graphs of crossing straight lines $L_{row}(x) = E^{i*}_{max}[t] - x \, (A'A'')$ and $L_{column}(x) = E^{*i}_{max}[t] + x \, (B'B'')$, with the resulting function $Z \, (A'CB')$ of the following type:

$$Z = max \, (L_{row}(x), \, L_{column}(x)) = \frac{L_{row}(x) + L_{column}(x)}{2} + \left| \frac{L_{row}(x) - L_{column}(x)}{2} \right| \tag{32}$$

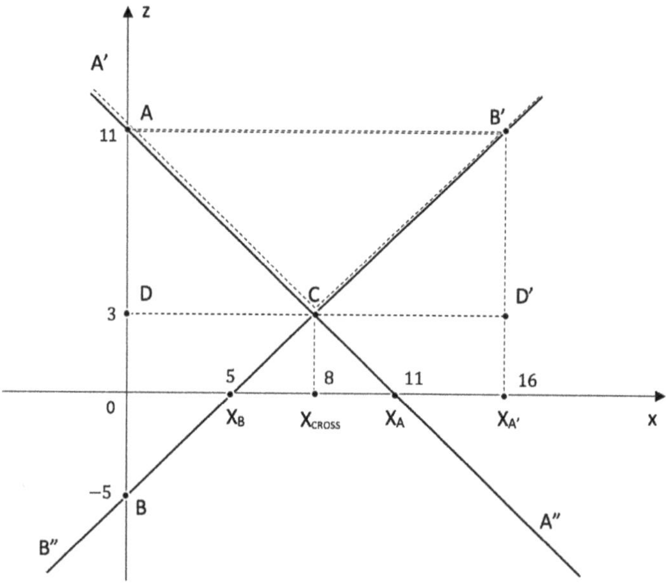

Figure 5. Geometric illustration of the vector compression method.

Moving to the point of intersection of the lines with the coordinate $X_{cross} = (E^{i*}_{max}[t] - E^{*i}_{max}[t])/2$, we reduce the modulus of the difference between $L_{row}(x)$ and $L_{column}(x)$.

Since the first addend does not depend on x, the result depends on the sign under the module. At the point of intersection of X_{cross}, the module is zero.

Transformation properties:

1. Zero neutrality: regardless of the type, zero displacement leaves the matrix unchanged

$$E[t+1] = S^i_\varnothing(E[t]) = E[t]. \tag{33}$$

2. Absorption: two transformations of the same type are added together

$$E[t+2] = S^i_{x2}(E[t+1]) = S^i_{x2}\left(S^i_{x1}(E[t])\right) = S^i_{x2+x1}(E[t]). \tag{34}$$

3. Rearrangement: two adjacent transformations, regardless of their types, can be rearranged

$$E[t+2] = S_{x2}^{i2}(E[t+1]) = S_{x2}^{i2}\left(S_{x1}^{i1}(E[t])\right) = S_{x1}^{i1}\left(S_{x2}^{i2}(E[t])\right). \tag{35}$$

4. Cyclicity: regardless of x, the following is executed

$$E[t+N] = S_x^1\left(S_x^2 \ldots \left(S_x^N(E[t])\right)\right) = E[t]. \tag{36}$$

Used to line up the final displacement parameters

$$E[t+2N] = S_{x1-x_N}^1\left(S_{x2-x_N}^2 \ldots \left(S_{\varnothing}^N(E[t])\right)\right). \tag{37}$$

5. Convolution of sequences by permutation and absorption:

$$E[t+\infty] = S_{x1+\ldots+x\infty}^1\left(S_{y1+\ldots+y\infty}^2 \ldots \left(S_{z1+\ldots+z\infty}^N(E[t])\right)\right) \tag{38}$$

is provided by permutation of transformations (by property 3) of the first type to the left with absorption (by property 2). The same procedures are then implemented for the second and subsequent types.

From properties 1–5, it follows that at any time t, the state of the matrix $E[t]$ is determined by the initial matrix $E[t_0]$ and accumulated sums for the same type of transformation, while the order of application of transformations of various types is not important.

4.2. Transordinate Vector Compression Method

The choice of displacing

$$\Delta L_i[t] = \frac{E_{max}^{i*}[t] - E_{max}^{*i}[t]}{2} = \frac{E_{max}^{i*}[t] + E_{min}^{i*}[t]}{2} \tag{39}$$

influences the matrix $E[t]$ with transformation $S_{\Delta L_i[t]}^i(E[t])$.

As a result, we get

$$E_{i,j}[t+1] = E_{i,j}[t] - \Delta L_i[t], \; if \; G_{i,j} = 1, \tag{40}$$

$$E_{j,i}[t+1] = E_{j,i}[t] + \Delta L_i[t], \; if \; G_{j,i} = 1. \tag{41}$$

The values $E_{max}^{i*}[t+1]$ and $E_{max}^{*i}[t+1]$ are aligned.

Definition 1. *The first norm $\|E\|$ is the maximum of the matrix $E_{i,j}[t]$ if $G_{i,j} = 1$.*

Two lemmas of convergence are true.

Lemma 1. *(by lines). If $E_{max}^{i*}[t] = \|E\|[t]$ is the only maximum, at which $E_{max}^{*i}[t] = \|E\|[t]$, then $\|E\|[t+1] = S_{\Delta L_{i*}[t]}^{i*}(\|E\|[t]) < E[t]$. (by line $i*$).*

Lemma 2. *(by columns). Let the single maximum be reached on element $E_{i*,j*}[t]$. Then, the decrease $\|E\|[t+1]$ is achieved due to the conversion $S_{\Delta L_{j*}}^{j*}(E[t])$ (by the column $j*$).*

Definition 2. *The second norm $[\![E]\!]$ is defined as the sum.*

$$[\![E]\!] = \frac{1}{2} \sum_{i=1}^{N} \left| E_{max}^{i*} + E_{min}^{i*} \right|. \tag{42}$$

By virtue of Lemmas 1 and 2, the process of lowering the first norms converges. If the first and second norms are zero, the matrix $E[t]$ becomes zero, and the matrix $\overline{A}$ becomes consistent and is fully determined by the value of the accumulated sums taken with the inverse sign.

For uncoordinated matrices $\overline{A}$, the convergence process described in Lemmas 1 and 2 results in final states other than the zero matrix $E[t]$. The second norm $[\![E]\!]$ becomes zero, and the first norm $\|E\|$ becomes equal to the value K, which in the future we will call the consistency criterion of the matrix $\overline{A}$.

Now, let us consider the set of lines $I = \{i\}$ for which E_{max}^{i*} is K. Each line $i_d \in I$ has at least one maximum $E_{max}^{i_d*}$ and at least one minimum $E_{min}^{i_d*}$. The remaining elements of matrix E for which $\left(-K < E_{i_dj} < K\right)$ is executed may be temporarily discarded.

We create oriented graph H_{ij} from $\left|E_{i_dj}\right| = K$. We will assume that the maxima are the entry points to the top i_d and the minima are the exit points from i_d. As a result, H_{ij} will have one or more cycles, and the cycles can be inserted into each other. The obtained graph is not necessarily connected, nor does it necessarily contain all the tops of the graph $\overline{a}_{ij}$.

Each cycle is similar to the known task in the theory of antagonistic games—the game "rock-paper-scissors", where the criterion of consistency K plays the role of a "bet" in one game. With a small bet, the game is quite harmless. The price of the game is zero, and there is a Nash balance—it is absolutely random, not subject to any algorithm and there is proportional use of all three strategies. With a large bet (not comparable to the smaller capital of one of the players), a "tragic" outcome is possible—no one will give credit to the loser to "recover his losses". How to reduce the game's bet? With large consistency criteria K, an unsolvable situation arises.

The first way out—as in the method of analyzing hierarchies, is to abandon the results of uncoordinated expertise.

The second way out is to abandon some grades from $E_{max}^{i_d*}$, and instead of $G_{ij} = 1$, accept $G_{ij} = 0$. It is necessary to break some kind of cycling connection, without destroying the connectivity of the remaining graph G_{ij}. By breaking the cycle with K, after recalculation v_i, there will be a smaller value $K_1 < K$.

The third way out is that communications remain the same: $G_{ij} = 1$ and new is not formed, but matrix weights $\overline{A}$ "recover", as specified in top set I. This is the best way to ensure it has counted K_1—the maximum value "preceding" the current value of K that does not participate in the construction of H_{ij}.

Let $0 < K_\varepsilon \le K_1 < K$ be executed for K_1. In fact, you can limit yourself to a non-zero minimum level of K_ε. Yet, when $K_\varepsilon < K_1$, matrix E must be subtracted from matrix $\overline{A}$, on which the mask from elements of a positive multiplier $\theta = K - K_1$ is imposed on elements from H_{ij}. Thereby, the new matrix of $\overline{A}$ will have the criterion of coherence K_1.

After that, it may be necessary to recalculate the "non-critical" (not included in the top set I) "accumulated sums" of lines and form a new matrix $H_{ij}(K_1)$. If this is for a new K_2, $K_\varepsilon < K_2$ will be executed. Let us continue the procedure as described above. If on any step $(\zeta = 1, \dots)$ $K_{\text{III}} < K_\varepsilon$, the step is summarizing and parameter θ decreases to $\theta_\phi = K_\zeta - K_\varepsilon$.

4.3. Gradient Vector Compression Method

The presence of two equivalent methods (line i and column j) serves as the basis for a gradient variant of realization of the vector compression method. As can be seen from the matrix transformation definition itself, optimal compression at each time point t is not necessary for success; it is only necessary to set the correct direction. To avoid searching each iteration for the "best pair" by the displacement value (Table 1), one may wish to reduce the displacement value several times.

Table 1. Data for calculating components of displacement values in the gradient method of vector compression.

No. of Line	Maximum Value in Line	Minimum Value in Line	Displacement Value
1	$E^{1*}_{max}[t]$	$E^{1*}_{min}[t]$	$(E^{1*}_{max}[t] + E^{1*}_{min}[t])/(2*Q)$
2	$E^{2*}_{max}[t]$	$E^{2*}_{min}[t]$	$(E^{2*}_{max}[t] + E^{2*}_{min}[t])/(2*Q)$
...			
N	$E^{N*}_{max}[t]$	$E^{N*}_{min}[t]$	$E^{N*}_{max}[t] + E^{N*}_{min}[t])/(2*Q)$

First, by computational experiment, and then, theoretically, it was possible to show for 3×3 matrices that an increase in the divider in the displacement value by a factor of $Q = 1.5$ turned out to be more effective. This allows the maxima and minima of the matrix to be recalculated once every N iterations.

As a result, the algorithm of the gradient method of vector compression is reduced to the following successive steps:

1. Calculate local maxima E^{i*}_{max} and minima E^{i*}_{min} by lines.

2. Recalculate $v^m_i[t+1] = v^m_i[t] + \frac{E^{i*}_{max} + E^{i*}_{min}}{3}$.

3. If $\frac{1}{2} \sum_{i=1}^{N} \left| E^{i*}_{max} + E^{i*}_{min} \right| > \varepsilon$, transition to item 1.

   Otherwise:

4. If $E^{**}_{max} > \varepsilon_1$ is a correction of $\overline{A} = \overline{A} - \theta E$, the previous v^m_i can be left.

   Transition to item 1.

   Otherwise:

   The specified accuracy has been achieved.

   In these steps, ε is a measure of the inaccuracy of the definition of a gradient of logarithms of weights, while ε_1 is the required accuracy of the decision.

   Obviously, inequities are being fulfilled

$$0 < \varepsilon \ll \varepsilon_1 \ll 1. \tag{43}$$

4.4. Hybrid Methods for Partial Pairwise Comparison with Fuzzy Information

The tool discussed in Section 4.3 is applicable for operation with one decision matrix (filled with half-significant elements [23]). To work with matrices of large dimensions, various methods of aggregating assessments are needed [34]. This is significantly observed in two extreme cases—when the number of experts M is large, and when the number of objects of comparison N is large. In either of these scenarios, a significant chunk of the resources will be spent on checking non-zero values of the indicator matrix $G_{i,j}$. Therefore, we propose representing generalized matrices in the form of "lists" of non-zero elements by lines $\{i\}$, or in the form of a multilayer regular neural network (NN), which implements the calculation of local maxima E^i_{max} and minima E^i_{min} by lines.

In this case, the calculation of "lists" can be regularized by the addition of maxima E^i_{max} and minima E^i_{min} into the functions of calculation to take the place of missing communication elements, such as communication with the first elements in the list (element in the first layer of the NN). They are marked in the examples below (in Table 2, grey on an orange background, and in Table 3, yellow on a grey background).

Table 2. Co-locating based on vector compression method.

	X1	X2	X3	X4	X5	X6	X7	X8	X9	X10	X11	s = 1		s = 2		s = 3		s = 4	
X1	NA	+0.1	NA	NA	NA	0	NA	NA	NA	NA	NA	2	+0.1	2	+0.1	6	+0.0	2	+0.1
X2	−0.1	NA	+0.3	NA	NA	NA	NA	0	NA	NA	NA	3	+0.3	1	−0.1	8	+0.0	3	+0.3
X3	NA	−0.3	NA	=0.2	NA	NA	NA	NA	NA	NA	NA	4	+0.2	2	−0.3	4	+0.2	4	+0.2
X4	0	NA	−0.2	NA	+0.1	NA	NA	NA	NA	0	NA	5	+0.1	3	−0.2	10	+0.0	5	+0.1
X5	0	0	NA	−0.1	NA	NA	NA	NA	NA	NA	NA	4	−0.1	4	−0.1	4	−0.1	4	−0.1
X6	0	NA	NA	NA	NA	NA	=0.2	NA	0	NA	NA	7	+0.2	7	+0.2	1	−0.0	9	+0.0
X7	NA	NA	NA	NA	NA	−0.2	NA	+0.3	NA	NA	NA	8	+0.3	6	−0.2	8	+0.3	8	+0.3
X8	NA	0	NA	NA	NA	NA	−0.3	NA	NA	NA	0	7	−0.3	7	−0.3	2	−0.0	11	+0.0
X9	NA	NA	NA	NA	NA	0	NA	NA	NA	+0.1	NA	10	+0.1	10	+0.1	6	−0.0	10	+0.1
X10	NA	NA	NA	0	NA	NA	NA	NA	−0.1	NA	+0.3	11	+0.3	9	−0.1	4	−0.0	11	+0.3
X11	NA	NA	NA	NA	NA	NA	NA	0	NA	−0.3	NA	10	−0.3	10	−0.3	8	−0.0	10	−0.3

Table 3. Group decision-making based on vector compression method.

	X1	X2	X3	X4	X5	X6	X7	X8	X9	X10	X11	X12	s = 1		s = 2		s = 3		s = 4	
X1	NA	+0.1	+0.3	NA	0.0	NA	NA	NA	0.0	NA	NA	NA	5	0.0	9	0.0	2	+0.1	3	+0.3
X2	−0.1	NA	NA	NA	NA	0.0	NA	NA	NA	0.0	NA	NA	6	0.0	10	0.0	1	−0.1	6	0.0
X3	−0.3	NA	NA	−0.2	NA	NA	0.0	NA	NA	NA	0.0	NA	7	0.0	11	0.0	1	−0.3	4	−0.2
X4	NA	NA	+0.2	NA	NA	NA	NA	0.0	NA	NA	NA	0.0	8	0.0	12	0.0	3	+0.2	8	0.0
X5	0.0	NA	NA	NA	NA	NA	+0.1	+0.3	0.0	NA	NA	NA	1	0.0	9	0.0	7	+0.1	8	+0.3
X6	Ta	0.0	NA	NA	NA	NA	−0.3	−0.4	NA	0.0	NA	NA	2	0.0	10	0.0	7	−0.3	8	−0.4
X7	NA	NA	0.0	NA	−0.1	+0.3	NA	NA	NA	NA	0.0	NA	3	0.0	11	0.0	5	−0.1	6	+0.3
X8	NA	NA	NA	0.0	−0.3	+0.4	NA	NA	NA	NA	NA	0.0	4	0.0	12	0.0	5	−0.3	6	+0.4
X9	0.0	NA	NA	NA	0.0	NA	NA	NA	NA	+0.2	NA	NA	1	0.0	5	0.0	10	+0.2	1	0.0
X10	NA	0.0	NA	NA	NA	0.0	NA	NA	−0.2	NA	−0.2	NA	2	0.0	6	0.0	9	−0.2	11	−0.2
X11	NA	NA	0.0	NA	NA	NA	0.0	NA	NA	+0.2	NA	−0.2	3	0.0	7	0.0	10	+0.2	12	−0.2
X12	NA	NA	NA	0.0	NA	NA	NA	0.0	NA	NA	+0.2	NA	4	0.0	8	0.0	11	+0.2	4	0.0

It is advisable for a group task to have a regular sub-table of inter-matrix links as the first links, in contrast, for co-scaling tasks, select positive elements straight above the main diagonal. All elements will be non-negative except for the last element in each scale. The actual extreme elements in the scales on the second layer are duplicated by the corresponding elements of the first layer.

So, the algorithm consists of the first layer and following layers of the same type. On the first layer, the first element of the "list matrix" is assigned

$$E^i_{max}[1] = -v_i + v_{j[1]} + \bar{a}_{ij[1]} \tag{44}$$

$$E^i_{min}[1] = -v_i + v_{j[1]} + \bar{a}_{ij[1]}. \tag{45}$$

On subsequent layers NN s = 2, . . . the following is carried out

$$E^i_{max}[s] = max\left\{ E^i_{max}[s-1]; -v_i + v_{j[s]} + \bar{a}_{ij[s]} \right\} \tag{46}$$

$$E^i_{min}[s] = min\left\{ E^i_{min}[s-1]; -v_i + v_{j[s]} + \bar{a}_{ij[s]} \right\} \tag{47}$$

It is clear that such an organization of calculations is beneficial when $M \times N$ is large. Examples of matrices recalculations for the full list (neural network) are given in Tables 2 and 3 (right side).

4.5. Group Decision-Making Based on Vector Compression Method

Let $O_1, O_2, \ldots, O_N$ now represent a set of comparison objects. Each expert $\Im_1, \Im_2, \ldots, \Im_M$ sets his own logarithmic matrix of pairwise comparisons $\left(\bar{a}^m_{ij}\right)$ and an indicator matrix $\left(\bar{G}^m_{ij}\right)$. The only condition is that the link graph forms the backbone graph [33].

We create a combining network in the form of a block matrix $\bar{A}$ of dimension $[M \times N, M \times N]$ (Table 3) and a zero vector $v^m_i[0]$ of dimension $[M \times N]$.

The difference between a new algorithm and the one described earlier is that corrections (item 4) are made only according to expert matrices.

So:

1. Calculate local maxima E_{max}^{i*} and minima E_{min}^{i*} by lines.

2. Recalculate $v_i^m[t+1] = v_i^m[t] + \frac{E_{max}^{i*} + E_{min}^{i*}}{3}$.

3. If $\frac{1}{2} \sum_{i=1}^{N} \left| E_{max}^{i*} + E_{min}^{i*} \right| > \varepsilon$, transition to item 1.

    Otherwise:

4. If $E_{max}^{**} > \varepsilon_1$ is a correction of all $\overline{A}^m = \overline{A}^m - \theta \overline{E}^m$, the former v_i^m can be left.

    Transition to item 1.

    Otherwise:

    The specified accuracy has been achieved.

6. Calculate v_i^m for each expert.

7. Make the top T_i and lower B_i assessments of sets of weights $B_i \leq v_i^m \leq T_i$.

Introducing the upper T_i and lower B_i assessments for each expert results in all or almost all of the weights in the "agreed group assessment" eventually coinciding in the sense of co-scaling. Although no final decision was received, the order of the various experts was upheld, at least for the most significant comparison objects. Here, "non-stationary methods" of the indicator matrix (with the removal of inconsistent links) [23] can be effectively used.

4.6. Method and Algorithm for Combining Scales of Conflicting and Incomplete Expert Judgments

Let there be $^{\langle q \rangle}X$—a positively defined, monotonically non-growing scale on which certain factors of influence are assessed ($q = 1, \ldots, Q$). Moving to the logarithmic scale, we get assessments

$$^{\langle q \rangle}_1 x = \log_{C_{\langle q \rangle}} \left(^{\langle q \rangle}_1 x \right) \geq ^{\langle q \rangle}_2 x \geq \ldots \geq ^{\langle q \rangle}_{\langle q \rangle n-1} x \geq ^{\langle q \rangle}_{\langle q \rangle n} x \geq ^{\langle q \rangle}_{\langle q \rangle n+1} x \geq \ldots \geq ^{\langle q \rangle}_{\langle q \rangle N} x, \tag{48}$$

where $^{\langle q \rangle}N$ is the total number of compared objects and $C_{\langle q \rangle}$ refers to the logarithmic bases.

The vector column of preferences $^{\langle q \rangle}_{\langle q \rangle n} a$ ($^{\langle q \rangle}n = 1, \ldots, ^{\langle q \rangle}N - 1$) is calculated using formula $^{\langle q \rangle}_{\langle q \rangle n} a = ^{\langle q \rangle}_{\langle q \rangle n} x - ^{\langle q \rangle}_{\langle q \rangle n+1} x$. For certainty, let us assume that $^{\langle q \rangle}_{\langle q \rangle n} a = 0$.

The column vector reflects the non-negative, by definition, upper-right diagonal of the matrix of links $^{\langle q \rangle}A$ of dimension $\left[^{\langle q \rangle}N, ^{\langle q \rangle}N \right]$. The maximum and minimum values of the line participating in the vector compression method [23,24], depending on $^{\langle q \rangle}N$, are built automatically via (44) and (45).

The minima and maxima take their final values only at the end of accounting for all inter-scale connections via (46) and (47).

The order of enumeration of many inter-scale links does not affect the total.

The areas of the preference column vector in which zero is observed, we call the areas of equality of objects. Thus, in the assessments, the preference column vector alternates between the areas of equality and those of strict positive inequality of ordered factors. Factors belonging to the same equality area can be repositioned relative to each other. The factors of strict positive inequality are not permutable; otherwise, the order of constructing the preference column vector will be destroyed.

Here, there is an interesting case of pairwise interaction of scales. Let us, based on some assumptions, determine the equality of factors between two groups. This may be physical equality, such as assessments of the factors of two different influence groups in which some factors are present in both groups (athletes participating in both competitions). Or it can be logical equality—for example, the factors $^{\langle q1 \rangle}_{\langle q1 \rangle n} x$ and $^{\langle q2 \rangle}_{\langle q2 \rangle m} x$ are considered equal if they have the same realization risk, for example, if they lead to the same market share losses.

Having two different scales can lead to contradictions. So, the statement

$$\left({}^{<q2>}_{<q2>m1}x > {}^{<q2>}_{<q2>m2}x\right)\&\left({}^{<q1>}_{<q1>n1}x > {}^{<q1>}_{<q1>n2}x\right)\& \\ \&\left({}^{<q2>}_{<q2>m1}x = {}^{<q1>}_{<q1>n2}x\right)\&\left({}^{<q1>}_{<q2>m2}x = {}^{<q1>}_{<q1>n1}x\right) \tag{49}$$

results in equality being recognized

$$\left({}^{<q2>}_{<q2>m1}x = {}^{<q2>}_{<q2>m2}x\right)\&\left({}^{<q1>}_{<q1>n1}x = {}^{<q1>}_{<q1>n2}x\right). \tag{50}$$

Both scales are compressed.

The same situation occurs in cases (51) and (52)

$$\left({}^{<q2>}_{<q2>m1}x > {}^{<q2>}_{<q2>m2}x\right)\&\left({}^{<q1>}_{<q1>n1}x = {}^{<q1>}_{<q1>n2}x\right)\&\left({}^{<q2>}_{<q2>m1}x = {}^{<q1>}_{<q1>n2}x\right)\&\left({}^{<q2>}_{<q2>m2}x = {}^{<q1>}_{<q1>n1}x\right) \tag{51}$$

$$\left({}^{<q2>}_{<q2>m1}x = {}^{<q2>}_{<q2>m2}x\right)\&\left({}^{<q1>}_{<q1>n1}x > {}^{<q1>}_{<q1>n2}x\right)\&\left({}^{<q2>}_{<q2>m1}x = {}^{<q1>}_{<q1>n2}x\right)\&\left({}^{<q2>}_{<q2>m2}x = {}^{<q1>}_{<q1>n1}x\right) \tag{52}$$

but with the compression of one of the two scales.

Thus, for a paired comparison of two scales, a general rule can be formulated: contradictions between two scales do not occur if and only if the sets of pairwise equations in the two scales are themselves an ordered set.

5. Discussion of Results

Let us analyze the typical examples.

Example 1. *Let us consider the interaction state of scales* ${}^{\langle1\rangle}X$ *and* ${}^{\langle2\rangle}X$ *in the form of the following matrix with inter-scale connections as corresponding equations (Table 4):* $\left({}^{\langle1\rangle}_{1}x = {}^{\langle2\rangle}_{1}x\right)$ $\&\left({}^{\langle1\rangle}_{2}x = {}^{\langle2\rangle}_{3}x\right)\&\left({}^{\langle1\rangle}_{3}x = {}^{\langle2\rangle}_{4}x\right)\&\left({}^{\langle1\rangle}_{4}x = {}^{\langle2\rangle}_{5}x\right)\&\left({}^{\langle1\rangle}_{5}x = {}^{\langle2\rangle}_{6}x\right).$

Table 4. Matrix of coefficients.

NA	0.1	NA	NA	NA	0.0	NA	NA	NA	NA	NA
−0.1	NA	0.3	NA	NA	NA	NA	0.0	NA	NA	NA
NA	−0.3	NA	0.2	NA	NA	NA	NA	0.0	NA	NA
NA	NA	−0.2	NA	0.1	NA	NA	NA	NA	0.0	NA
NA	NA	NA	−0.1	NA	NA	NA	NA	NA	NA	0.0
0.0	NA	NA	NA	NA	NA	0.2	NA	NA	NA	NA
NA	NA	NA	NA	NA	−0.2	NA	0.3	NA	NA	NA
NA	0.0	NA	NA	NA	NA	−0.3	NA	0.1	NA	NA
NA	NA	0.0	NA	NA	NA	NA	−0.1	NA	0.1	NA
NA	NA	NA	0.0	NA	NA	NA	NA	−0.1	NA	0.3
NA	NA	NA	NA	0.0	NA	NA	NA	NA	−0.3	NA

It can be seen that both scales are consistent according to the general formed rule. A directed graph is similar to a joint network of two performers in which there are restrictions in the form of information links. This allows us to calculate the late completion of "work" in each node and the corresponding reserves of "work" (Table 5).

Table 5. Given data for optimal time reserve solution.

Index	${}^{\langle1\rangle}X$	${}^{\langle1\rangle}a$	${}^{\langle1\rangle}r$	${}^{\langle1\rangle}OptX$	${}^{\langle2\rangle}X$	${}^{\langle2\rangle}a$	${}^{\langle2\rangle}r$	${}^{\langle2\rangle}OptX$
1	0.7	0.1	0.4	1.3	1.0	0.2	0	1.3
2	0.6	0.3	0	0.8	0.8	0.3	0	1.1
3	0.3	0.2	0	0.5	0.5	0.1	0.2	0.8
4	0.1	0.1	0.2	0.3	0.4	0.1	0.1	0.5
5	0	0	0	0	0.3	0.3	0	0.3
6					0	0	0	0

It can also be seen that time stretches affect both scales. We get an extended scale that, in particular, is piecewise-continuous. This is quite natural since private risk-rating is often measured in pieces. When the optimal time $T_{opt} = 1,3$ (Table 6) does not suit—for example, we cannot wait long for the end of all work $j^{\langle final\rangle}a$—we must solve the task of partially compressing some "critical work" in both scales.

Table 6. Dynamics of reduction of time reserves from the value of total time allocated for the complete scope of work.

	SC	T_{opt}				T^*_{opt}					
$X_1^{\langle 1\rangle}$	0.7	1.30	1.200	1.000	0.9(3)	0.9000	0.80	0.7000	0.40	0.100	0.0
$X_1^{\langle 2\rangle}$	1.0	1.30	1.200	1.000	0.9(3)	0.9000	0.80	0.7000	0.40	0.100	0.0
$X_2^{\langle 2\rangle}$	0.8	1.10	1.100	0.925	0.8(6)	0.8375	0.75	0.6625	0.40	0.100	0.0
$X_2^{\langle 1\rangle}$	0.6	0.80	0.900	0.750	0.7(0)	0.6750	0.60	0.5250	0.30	0.075	0.0
$X_3^{\langle 2\rangle}$	0.5	0.80	0.900	0.750	0.7(0)	0.6750	0.60	0.5250	0.30	0.075	0.0
$X_3^{\langle 1\rangle}$	0.3	0.50	0.600	0.500	0.4(6)	0.4500	0.40	0.3500	0.20	0.050	0.0
$X_4^{\langle 2\rangle}$	0.4	0.50	0.600	0.500	0.4(6)	0.4500	0.40	0.3500	0.20	0.050	0.0
$X_4^{\langle 1\rangle}$	0.1	0.30	0.300	0.250	0.2(3)	0.2250	0.20	0.1750	0.10	0.025	0.0
$X_5^{\langle 2\rangle}$	0.3	0.30	0.300	0.250	0.2(3)	0.2250	0.20	0.1750	0.10	0.025	0.0
$X_5^{\langle 1\rangle}$	0.0	0.00	0.000	0.000	0.0(0)	0.0000	0.00	0.0000	0.00	0.000	0.0
$X_6^{\langle 2\rangle}$	0.0	0.00	0.000	0.000	0.0(0)	0.0000	0.00	0.0000	0.00	0.000	0.0

This is similar to the application of additional resources (human and material) to reduce the risks of the entire project (e.g., time delays). In non-critical locations, compression will be carried out by reducing the time reserve for performing non-critical work, until the reserve for such work runs out and the work becomes critical. The yellow color indicates when the work ceases to have a reserve and becomes critical as T_{plan} decreases.

If $T^*_{opt} = 0,9(3)$, a second optimal solution will be achieved, in which the maximum time reserve is equal to the maximum compression.

The solution of T^*_{opt} is interesting because it is achieved without collapsing the work on the general scale.

By collapse, we mean the situation where the private scale nodes offered for this work cease to be distinguishable. In this example, the first and only collapse occurs when $T_{plan} = 4$ with work $1^{\langle 2\rangle}a$. Time $T_{plan} = 4 < T^*_{opt}$, but there are often cases where works (scale nodes, objects) are located close to one another already on a private scale and collapse can thus occur ahead of the balance time.

Finally, if $T_{plan} = 0$, all scales collapse, which indicates that the objects of comparison are not distinguishable on the generalized scale. If we have a lot of resources, then they can be spent on eliminating all the disadvantages $\left(T_{plan} \to 0\right)$ and the initial state of objects in both scales becomes insignificant.

Thus, both the remoteness of the target and the available resource affect the overall risk.

Example 2. *Let us now consider the solution of summarizing private assessments of the order of preference of groups of different types of objects, as ranked within their groups into a single "group" preference. We solved a similar task in determining the systemic significance of different types of objects of a gas transmission system (compressor stations, gas distribution systems, underground gas storage facilities, etc.) [34].*

Based on physical and organizational principles, we add two dummy objects to these objects for each object type: TOP and BOTTOM. The TOP of type m is assigned the maximum achievable values that objects of type m could reach on each measured scale, while the BOTTOM of type m is assigned the minimum achievable values that objects of type m could reach on each measurement scale under consideration. That is, the BOTTOM and TOP values are pre-calculated boundaries of the possible change in assessments of all objects on the corresponding measurement scale. Down-

loadable assessments for each data type are obtained with their own relative error, which makes it possible to determine the extremum maximum size of the cluster in the corresponding types of measurement scales.

The accuracy of assessing an object on a scale depends both on the size of the cluster and on the number of objects of the base type that correspond to the scale, and through the values of the assessments based on which other types of objects are recalculated. If in some scales the error is large or there are few objects, then it is obvious that the assessments of all objects on this scale will garner less "trust" than more accurate assessments for more objects. Next, the assessments of objects are "logarithmed" (in a conditional example, the decimal logarithms of the assessments are considered).

Like group expertise, in which matrix inconsistency is a rule rather than an exception and which leads to cyclic periodic closures, the presence of a cycle ($X1 = Z3$) & ($Z3 < Z2$) & ($Z2 = Y4$) & ($Y4 < Y1$) & ($Y1 = X3$) & ($X3 < X1$) leads to clusters in the solution due to inconsistency of the original matrices. Scales are inconsistent if there is almost no ordered set of their elements. All objects on all private scales are a partially ordered set. In the previous Example 5.1, clusters were formed before a full merger due to the uniformity of risks.

The meaning of the procedure for combining different types of objects into one list in the present example is the synthesis of the integral risk for each object, taking into account its own significance. We are looking for a solution that leads to minimal formation of clusters (mergers of objects) on the general scale. Optimal clustering is essentially the optimal risk of the entire design given the actual material.

The organization of the procedure for assigning an assessment to experts is an independent task. Experts are not required to know either the order of the types that are selected for examination or the values of intra-system indices. It is beneficial to know the results of object comparisons given by other experts. The initial data are a list of "approximate equations" (fuzzy equivalences) compiled by experts. It does not matter that there are no compared objects, and some objects can be present several times. A balanced solution is shown in Figure 6.

Eight clusters were formed, but only two had the same types of objects. This suggests that persistence was partially violated in them, which may indicate a "lie detection" [35]:

- *errors in the initial data;*
- *discovery of new "unique" properties that certain scales did not have separately—"emergence".*

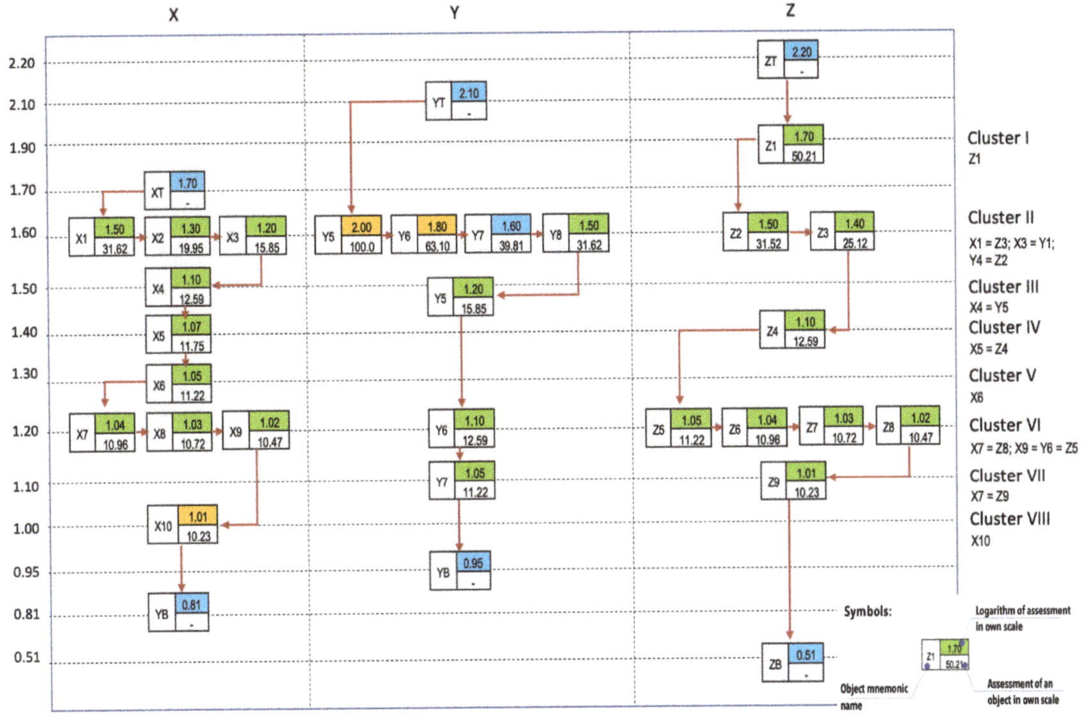

Figure 6. Building a balanced solution to the task of making an integral assessment of different types of objects.

6. Conclusions

This article analyzed the foundations of the future methodology for synthesizing the risk of functional smart expansive systems, taking into account the need to consider the balance between its constituent subsystems: production, development potential and regulation. We propose considering risks as "development anti-potentials" that slow down the reproduction speeds of the entire system. The concept of the geometric integral "anti-risk" is introduced, resulting from the concept of "difficulties in achieving the goal". Thus, the conceptual definition of risk as the influence of uncertainties on the achievement of the goal of smart expansive systems is formalized.

To assess private risk factors included in the integral risk, we propose a method of vector compression. The idea to build compatible reference solutions, which form the basis for the developed method, represents an alternative to pairwise comparison in the method of analysis of hierarchies and the method of analytical networks.

Further to this, we propose an approach to processing partial matrices of pairwise comparisons, which makes it possible to minimize the disadvantages of the existing methods for working with similar matrices, especially for matrices of large dimensions. The principles of handling pairwise comparison matrices by describing their upper and lower boundaries have been investigated. The developed vector compression algorithm allows us to obtain the weights of compared objects on the basis of matrices of pairwise comparisons containing omissions, without fully restoring the matrix of pairwise comparisons, and also allows us to obtain the weights of given upper and lower boundaries through comparative assessment of pairs of objects.

This paper is not a standalone work capable of covering all issues and presenting the variety of smart expansive systems. These will undoubtedly be the topics of further research. To give an example of where future studies may take us, the smart expansive system in this paper was considered in a linear approximation and stationary case. Beyond the scope of the article, a question must also be raised about the behavior of an intelligent expansive system in a "non-stationary state", where oscillatory processes (and maybe chaos) may occur. Moreover, the uniqueness of the approximation we chose has not been proven, and the option of using the vector compression method for upper and lower bounds in cases of restrictions imposed on the coefficient (greater than/less than zero) for fuzzy definition of the original matrices has not been considered. We plan to investigate all these and many other avenues in the future.

The proposed method could become an important element in the algorithmic provision of expert advising systems to support decision-making on the management of smart expansive systems, provided there is an appropriately organized procedure for selecting experts to be involved in the assessment of solutions.

Author Contributions: Conceptualization, A.B. and N.Z.; methodology, A.B.; software, A.R.; validation, A.B., N.Z. and N.K.; formal analysis, N.K.; investigation, N.Z.; data curation, N.Z.; writing—original draft preparation, A.B.; writing—review and editing, A.B.; visualization, A.R.; supervision, N.Z.; project administration, A.B. All authors have read and agreed to the published version of the manuscript.

Funding: This research received no external funding.

Conflicts of Interest: The authors declare no conflict of interest.

References

1. Petukhov, G.B. *Fundamentals of Efficiency Theory of Targeted Processes. Part 1. Methodology, Methods, Models*; Publishing House of Ministry of Defense of the USSR: Moscow, Russia, 1989; 647p.
2. Kantorovich, L.V. *Mathematical and Economical Works*; Selectas., M., Ed.; Nauka: Moscow, Russia, 2011; 760p.
3. Wosserman, P. *Neural Computing: Theory and Practice*; Mir: Moscow, Russia, 1992; 240p, (Translated from English).
4. Kohonen, T. *Self-Organizing Maps*; Springer: New York, NY, USA, 1995.
5. Specht, D. Probabilistic Neural Networks. *Neural Netw.* **1990**, *3*, 109–118. [CrossRef]
6. Vyugin, V. *Mathematical Principles of Machine Learning and Forecasting Theory*; MCMCE: 2013; 390p. Available online: http://iitp.ru/upload/publications/7138/vyugin1_eng.pdf (accessed on 13 December 2021).
7. Gladkov, L.A.; Kureichik, V.V.; Kureichik, V.M. *Genetic Algorithms: Study Guide*, 2nd ed.; Physmathlit: Moscow, Russia, 2006; 320p.
8. Taha, H.A. *Operations Research*; Williams: London, UK, 2016; 912p.
9. Germayer, Y.B. *Introduction to Operations Research Theory*; Nauka: Moscow, Russia, 1971; 383p.
10. Saridis, J. *Self-Organizing Stochastic Management Systems*; Nauka: Moscow, Russia, 1980; 397p.
11. Kalmykov, S.A.; Shokin, Y.I.; Yuldashev, Z.K. *Methods of Interval Analysis*; Nauka: Novosibirsk, Russia, 1986; 224p.
12. Zade, L.A. *Fuzzy Sets and Their Application in Forms Recognition and Cluster Analysis—Classification and Cluster*; Van Raizin, J., Ed.; Mir: Moscow, Russia, 1980; pp. 208–247.
13. Yager, R.R. (Ed.) *Fuzzy Sets and Possibility Theory. Recent Achievements*; Radio and Svyaz: Moscow, Russia, 1986; 408p, (Translated from English).
14. Larichev, O.I. *Theory and Methods of Decision Making and Chronicle of Events in Magical Countries*, 3rd ed.; Logos: Newberry, FL, USA, 2006; 392p.
15. Fodor, J.; Roubens, M. *Fuzzy Preference Modelling and Multicriteria Decision Support*; Springer: Dordrecht, The Netherlands, 1994. [CrossRef]
16. Larichev, O.I.; Petrovsky, A.B. *Decision Making Support Systems. Current State and Prospects for Their Developments—Results of Science and Technics*; Series—Technique Cybernetics V.21; VINITI: Moscow, Russia, 1987; pp. 131–164. Available online: http://www.raai.org/library/papers/Larichev/Larichev_Petrovsky_1987.pdf (accessed on 13 December 2021).
17. Saaty, T. *Decision Making. Hierarchy Analysis Method*; Vachnadze, R.G., Translator; Radio and Svyaz: Moscow, Russia, 1993; 278p, (Translated from English).
18. Nogin, V.D. Simplified Version of Hierarchy Analysis Method based on nonlinear convolution of criteria. *Mag. Numer. Math. Math. Phys.* **2004**, *7*, 1261–1270.
19. Podinovsky, V.V.; Podinovskaya, O.V. About Incorrect Hierarchy Analysis Method. *Manag. Probl.* **2011**, *1*, 8–13.
20. Gusev, S.S. Analysis of Methods and Approaches for Solving Tasks of Multi-Criteria Choice in Conditions of Uncertainty. *Interact. Sci.* **2018**, *1*, 69–75. [CrossRef]
21. Saaty, T.L. *Decision Making in Dependencies and Feedback: Analytical Networks*; Book House «Librocom»: Moscow, Russia, 2009; 360p.

22. Seredkin, K.A. On the Limits of Applicability of the Method of Analytical Networks in Tasks of Decision Making in Natural Sciences. *Artif. Brain Decis. Mak.* **2018**, *2*, 95–102.

23. Bochkov, A.V.; Zhigirev, N.N.; Ridley, A.N. Method of recovery of priority vector for alternatives under uncertainty or incomplete expert assessment. *Dependability* **2017**, *17*, 41–48. [CrossRef]

24. Ridley, A.N. Risk synthesis methodology in system management. In *Gagarin's Reading—2019 Collection of Abstracts of XLV Reports of the International Youth Scientific Conference*; Moscow Aviation Institute National Research University: Moscow, Russia, 2019.

25. Glushkov, V.M.; Ivanov, V.V.; Yanenko, V.M. *Modelling of Developing Systems*; Nauka: Moscow, Russia, 1983; 350p.

26. Glushkov, V.M. *Introduction to the Theory of Self-Perfecting Systems*, 2nd ed.; Carryover of Academician V.M. Glushkov—Sciences on Artificial, No. 42; LENAND: Tallahassee, FL, USA, 2022; 112p.

27. Onyky, B.N.; Erivansky, Y.E.; Semenov, L.L.; Mikhailov, D.V. Automated Scientific and Production Association Management System (AMS «Extremum»): Technical Project; 17th Main Branch; Book 4: Subsystem of Techno-Economic Production Planning; R&D Technical Report, Moscow, Russia, 1974; 113p. Available online: https://www.ametsoc.org/index.cfm/ams/get-involved/ (accessed on 13 December 2021).

28. Onyky, B.N.; Erivansky, Y.E.; Reznichenko, V.Y. Automated Scientific and Production Association Management System (AMS « Extremum»): Technical Project; 17th Main Branch; Book 2: Scientific Research and Development Management Subsystem; R&D Technical Report, Moscow, Russia, 1974; 124p. Available online: https://www.ametsoc.org/index.cfm/ams/get-involved/ (accessed on 13 December 2021).

29. Zhigirev, N.N. Man-Machine Procedures for Resource Allocation in Developing Systems. Ph.D. Thesis, National Research Nuclear University, Moscow, Russia, 1987.

30. Prigogine, I.; Lefever, R. Symmetry Breaking Instabilities in Dissipative Systems II. *J. Chem. Phys.* **1968**, *48*, 1695–1700. [CrossRef]

31. Kaplinsky, A.I.; Russman, I.B.; Umyvakin, V.M. *Algorithmization and Modeling of Weakly Formalized Tasks for Selecting the Best System Options*; Publishing House VSU: Voronezh, Russia, 1991; 168p.

32. Arnold, V.I. *Catastrophe Theory*, 3rd ed.; Amended; Nauka: Moscow, Russia, 1990; pp. 31–32.

33. Bochkov, A.; Niias, J.; Ridley, A.; Kuzmina, N.; Zhigirev, N. Vector compression method to convert the incomplete matrix of pairwise comparisons in the analytic hierarchy process. In Proceedings of the International Symposium on the Analytic Hierarchy Process, Web Conference, 3–6 December 2020. [CrossRef]

34. Bochkov, A.; Lesnykh, V.; Zhigirev, N.; Lavrukhin, Y. Some methodical aspects of critical infrastructure protection. *Saf. Sci.* **2015**, *79*, 229–242. [CrossRef]

35. Gorbatov, V.A. *Fundamental Foundations of Discrete Mathematics*; Physmathlit; Nauka: Moscow, Russia, 1999; 544p.

 *mathematics*

Article

New Applied Problems in the Theory of Acyclic Digraphs

Gurami Tsitsiashvili [1,*] and Victor Bulgakov [2]

[1] Institute for Applied Mathematics, Far Eastern Branch of Russian Academy of Sciences, 690041 Vladivostok, Russia

[2] Federal Scientific Center of the East Asia Biodiversity, Far Eastern Branch of Russian Academy of Sciences, 690022 Vladivostok, Russia; bulgakov@biosoil.ru

* Correspondence: guram@iam.dvo.ru; Tel.: +7-914-693-2749

Abstract: The following two optimization problems on acyclic digraph analysis are solved. The first of them consists of determining the minimum (in terms of volume) set of arcs, the removal of which from an acyclic digraph breaks all *paths passing through* a subset of its vertices. The second problem is to determine the smallest set of arcs, the introduction of which into an acyclic digraph turns it into a strongly connected one. The first problem was solved by reduction to the problem of the maximum flow and the minimum section. The second challenge was solved by calculating the minimum number of input arcs and determining the smallest set of input arcs in terms of the minimum arc coverage of an acyclic digraph. The solution of these problems extends to an arbitrary digraph by isolating the components of cyclic equivalence in it and the arcs between them.

Keywords: acyclic digraph; maximal flow; minimal cut; minimal arc cover; bipartite digraph

Citation: Tsitsiashvili, G.; Bulgakov, V. New Applied Problems in the Theory of Acyclic Digraphs. *Mathematics* 2022, 10, 45. https://doi.org/10.3390/math10010045

Academic Editor: Frank Werner

Received: 16 November 2021
Accepted: 22 December 2021
Published: 23 December 2021

Publisher's Note: MDPI stays neutral with regard to jurisdictional claims in published maps and institutional affiliations.

1. Introduction

The monographs [1–3], which have become classic, are devoted to theoretical and applied issues of digraph research. They are closely related to the Ford–Fulkerson theorem on the equality of the maximum flow and the minimum cut [4,5]. At first glance, many problems with digraphs look like NP-problems. However, with a special selection of optimized indicators and graph transformation, these tasks can be reduced to the search for Ford–Fulkerson algorithms. In this way, it becomes possible to avoid the appearance of NP-problems when working with digraphs. Therefore, these studies can be attributed to the intensively developing applications of digraph theory in system analysis and the theory of optimization algorithms on graphs [6,7]. The papers closest to the subject of this article can be considered [8,9].

In particular, the strong connectivity in digraphs is considered in the presence of arc failures [10]. The paper [11] studies the use of digraphs in interferometry. The paper [12] explores the issues of spectral complexity of digraphs and their application to structural decomposition. In [13], using digraph models, the issues of signal processing and learning from network data are analysed. In the work [14], the multilevel task of identifying bottlenecks in the network is considered. In the works [15,16], minimal networks are built in which the Ford–Fulkerson procedure may not be completed. The paper [17] explores various ways of applying stochastic models on digraphs in computational biology. Due to the intensive development of biotechnology, two new applied problems of digraph theory are posed in this paper, which require both the development of the theory and the construction of new optimization algorithms on digraphs. Both tasks were initiated by biotechnological problems related to protein networks. Let us describe these tasks in more detail.

The importance of analysing protein networks in plant bioengineering is due to the fact that the growth and development of plants, as well as their protective functions, are regulated by the interaction of various protein signalling modules. At the same time, fine-tuning of metabolic processes takes place, allowing the plant to adapt to changing environmental conditions. Plant interactomes have not been worked out enough yet. Therefore, it is necessary to develop mathematical modelling methods that describe the natural functions of protein modules as accurately as possible.

Previously, we tried to link various cellular processes affecting the biosynthesis of anthocyanins. In this work, we have identified the main signalling protein modules that regulate the biosynthesis of anthocyanins. As far as we know, this was the first reconstruction of a network of proteins involved in the secondary metabolism of plants [18].

It was further shown that the signalling systems of abscisic acid (ABA) and chaperones are integrated by chromatin remodelling proteins (CRC) into a single regulatory network. CRC proteins "remember" the previous stressful effects and adjust the plant to the perception of new ones, and memory is generated in the offspring. A new scientific direction—"bioengineering of memory" was substantiated [19].

Fabregas et al. [20] reported that overexpression of the vascular brassinosteroid receptor BRL 3 provides improved drought resistance without disrupting plant growth. We have constructed a network of protein–protein interactions of ABA and brassinosteroid signalling systems. It has been established that the phenomenon of drought resistance mediated by BRL3 can be explained by the generation of stress memory (a process known as "priming" or "acclimation") [21]. Let us now turn to the mathematical formulation and algorithmic solution of the protein network modelling.

In the first problem, we are talking about finding a smallest set of arcs, the removal of which blocks all paths passing through an acyclic digraph with a set of W (corrupted) vertices. To solve this problem, it is proposed to cut the pathways entering the set of affected proteins or leaving this set, thus minimizing the number of cut paths. Such minimization deforms the structure of the protein network in the least way. It should be noted that isolation measures restricting the normal functioning of various communication networks, such as transport, economic, educational, etc., have recently invaded our lives. Therefore, the considered biotechnological problem acquires a more general meaning, which requires the construction of economical algorithms for its solution.

In biotechnology there is a problem to decrease a number of blocking arcs. To solve this problem it is possible to add to incoming and out coming arcs some arcs between corrupted vertices (selected by the biotechnologists) and to choose among them minimal number of blocking arcs. This procedure may be realized by implementing a large bandwidth to chosen arcs. All its stages are well known, but together they make it possible to solve an important and new problem in the field of biotechnology. Thus, it becomes possible to structure both the formulation and the solution of this problem, taking into account the choice of biotechnologists.

The solution of this problem is based on a special building of the integer bandwidths of the arcs of an acyclic digraph, the minimum section in which contains only of arcs entering from the inside into the subset W and/or exiting from the inside of the subset W and some arcs between vertices from W. By choosing the integer bandwidths, the arcs connecting the vertices of the subset U_*, are made unsaturated by the maximum flow. In turn, saturated with the maximum flow (the maximum flow passing through them coincides with their throughput) can only be arcs entering from the outside into a subset of W and/or exiting from the outside of the subset W. Next, the minimal section is searched using the well-known Ford–Fulkerson algorithm [4,5], which guaranteed to converge only for integer throughput. This algorithm was developed in its modifications [22,23]. Minimizing the number of arcs satisfying certain properties creates a risk of encountering an NP-problem as the problem of continuous brute force. The use of techniques that lead the tasks to the modified Ford–Fulkerson algorithm allows us to avoid the risk of NP-complex problems. Thus, in order to use the Ford–Fulkerson algorithm in solving this problem, it is necessary to build a two-pole and select the integer throughputs of its arcs so that the solution obtained in this algorithm determines the solution of the problem.

An alternative and in some sense inverse second problem is connected with the introduction into the digraph of a smallest set of new arcs that turn an acyclic digraph into a strongly connected one (in which there is a path from any vertex to any other vertex). This procedure is needed to include all the vertices of the acyclic digraph in the feedbacks that stabilize the functioning of the network represented by the digraph.

To do this, we first consider a bipartite acyclic digraph, in which all arcs are directed from the first lobe to the second. By removing the orientation of the arcs, we obtain an undirected bipartite graph. Using the well-known generalizations of the Ford–Fulkerson [24] algorithm, the maximum matching is searched for in it by a method of increasing alternating paths [25], from which the minimum arc cover is constructed. It consists of an incoherent collection of star-like sub graphs (in which arcs connect some vertices with the base of the star) ([26] p. 318). Next, the orientation of the arcs is restored and additional arcs are introduced in the resulting bipartite digraph by a special algorithm. This algorithm is based on the sequential arrangement first of the stars of the first type-with roots in the first lobe and then of the stars of the second type with roots in the second lobe. If all the stars have the same type, then their vertices may be connected by a Hamiltonian cycle [27].

However, if there are stars of different types, then the stars of the first and then the second type are located first. Then, additional arcs are introduced from the vertices of the second lobe to the vertices of the first lobe so that their number equals the maximum between the number of vertices of the first lobe and the second lobe. This number of additional arcs is minimal for obtaining strongly connected digraph. When moving from the minimum arc cover to the original bipartite digraph, the minimum number of additional arcs cannot increase. At the same time, the set of additional arcs already found transforms this bipartite digraph into a strongly connected one. The transition from an arbitrary acyclic digraph to a bipartite one is based on the allocation in an arbitrary digraph of the first lobe, including vertices from which only arcs come out, and the second lobe, including vertices that only arcs enter. An arc between the vertex of the first lobe to the vertex of the second lobe in a bipartite digraph is drawn if and only if there is a path between these vertices in the original digraph.

When solving both problems, we have to deal with digraphs that are not initially acyclic. The transformation of an arbitrary digraph into an acyclic one is based on the procedure for allocating cyclic equivalence classes (in which there is a path from any vertex to any other vertex) and arcs between them [28]. The paper presents an original algorithm for solving the problem of allocating cyclic equivalence classes [29] basing on the sequential inclusion in the digraph of a new vertex and arcs connecting it to the already specified ones.

2. Optimal Blocking of Selected Vertices of the Acyclic Digraph

Consider an acyclic digraph $\mathcal{G}$, with a finite set of vertices $\mathcal{U}$ and a finite set of arcs $\mathcal{V}$. In the set $\mathcal{U}$, a subset $U \subset \mathcal{U}$ of the so-called corrupted vertices is allocated. Let's define an acyclic digraph $G \subset \mathcal{G}$, with a set of vertices U and a set of arcs V connecting these vertices.

We attach to the digraph G the arcs of the digraph $\mathcal{G}$, walking to the set U from the set $\mathcal{U} \setminus U$ and the arcs coming out of U to $\mathcal{U} \setminus U$ and some arcs from the set $\mathcal{U}$, selected by biotechnologists. The set of these arcs is denoted by W and we call the *path passing through* the set U in the digraph $\mathcal{G}$, if it starts at the vertex of the set $\mathcal{U} \setminus U$ moves by an arc of W to the set U, passes through the set U and then moves by an arc of the set W to the set $\mathcal{U} \setminus U$. Our task is to determine in the set W the smallest subset of arcs whose removal from the digraph $\mathcal{G}$ breaks all paths, passing through the set of vertices U.

To do this, we transform the digraph G together with the set of its incoming and out coming arcs into the digraph G' as follows. All vertices of the set $\mathcal{U} \setminus U$, from which arcs of the set W move to U we combine into one vertex S and call it the *source*. All vertices of the set $\mathcal{U} \setminus U$, to which arcs of the set W move from U we combine into one vertex T and call it a *drain*. All arcs of the set W, going from the source S to the vertex $P \in U$, we combined into one arc w_* and determine its throughput equal to $n(P)$, which is the number of arcs to be combined. All arcs of the set W, going from the vertex $P \in U$ to the drain T, we combine into one arc w^* and determine its throughput equal to $N(P)$. If there is not an arc from the source S to vertex P, then $n(P) = 0$, if there is not an arc from the vertex P to the drain T, then $N(P) = 0$ as well.

Note that to formulate the optimization problem, it is necessary to exclude from its solution all arcs of the set V, which are not chosen by biotechnologists. To realize this procedure denote $L = \min\left(\sum_{P \in U} n(P), \sum_{P \in U} N(P) \right)$ and assign the bandwidth $L + 1$ to all arcs of the digraph G', which are not included in the set W. All arcs of the digraph G', which are chosen by biotechnologists, receive bandwidth 1. Therefore, the acyclic digraph G' becomes the bipolar, that is the digraph with the single vertex S, which has only out coming arcs, and the single vertex T, which has only incoming arcs.

Proposition 1. *The quantity of maximal flow in the bipolar G' is not larger than L.*

Proof of Proposition 1. Indeed, the quantity of maximal flow in the bipolar G' is not larger than a sum $\sum_{P \in U} n(P)$ of weights of arcs moving from the source S. The quantity of maximal flow in the bipolar G' is not larger than a sum $\sum_{P \in U} N(P)$ of weights of arcs moving from the drain T. Consequently, the quantity of the maximal flow in the bipolar G' is not larger than $L = \min\left(\sum_{P \in U} n(P), \sum_{P \in U} N(P) \right)$. $\square$

Proposition 2. *If in bipolar G', some arc w has bandwidth $L + 1$, then it does not include into any minimal cut.*

Proof of Proposition 2. From the theorem of Ford–Falkerson [4,5], it is clear that any minimal cut in the bipolar G' consists only of arcs saturated by any maximal flow. Therefore, the arcs which have bandwidths $L + 1$ cannot be included into any minimal cut. $\square$

Remark 1. *It follows from Propositions 1 and 2 that the proposed method for setting the bandwidths in the two-pole G' allows determining the minimum cuts only from the arcs of the set W, as required in the original formulation of the problem.*

As all included bandwidths are integers, using the Ford–Fulkerson algorithm [4,5] (or its modifications [22,23]) it is possible to calculate the maximum flow in the digraph G' and to obtain minimal cut W'. Each of the arcs of the set W', going from S to $P \in U$ is defined as the union of $n(P)$ arcs of the digraph $\mathcal{G}$. Similarly, each of the arcs of the set W', going from $P \in U$ to T is defined as the union of $N(P)$ arcs of the digraph $\mathcal{G}$. Denote W'' a set of arcs from the set W, included in the combined arcs of the set W'. Consequently from Propositions 1 and 2 and Remark 1 we have that W'' is the solution of the optimization problem of selecting a smallest set of arcs from the set W whose removal breaks all paths passing through the vertex set of U. Indeed, if the united arc $w \in W'$, then to block all paths passing through the vertex set U, we must delete all arcs, united in an arc w.

3. Optimal Algorithm for Converting an Acyclic Digraph into a Strongly Connected One

Problem statement. Suppose that a complex system, for example, a protein network, is represented by an acyclic digraph $\mathcal{G}$ without loops and isolated vertices. Let's denote V_1 the set of vertices from which the arcs only come out, and V_2 the set of vertices into which the arcs only enter. Now let's construct a bipartite digraph G, in which the set of vertices of the first lobe V_1, and the set of vertices of the second lobe V_2, the vertex $v_1 \in V_1$ is connected to the vertex $v_2 \in V_2$ by an arc, if there is a path between them in the acyclic digraph $\mathcal{G}$. As the digraph $\mathcal{G}$ is acyclic so there are not ways from vertices of the second lobe V_2 to vertices of the first lobe V_1. We denote $p(\mathcal{G})$ the smallest set (by a number of arcs) of additional arcs (call them "good arcs"), the introduction of which in $\mathcal{G}$ transforms it into a strongly connected digraph and designate $|p(\mathcal{G})|$ the number of good arcs in $p(\mathcal{G})$.

Let us transform a bipartite digraph G into an undirected one by removing the orientation of the arcs and find a minimal arc cover in it (see, for example, [6,24]). To do this, using the algorithm of increasing alternating paths, we find the maximum matching that can be transformed into a minimum arc cover, whose connected components are star-like sub graphs (all arcs come from one vertex or enter one vertex, called the root). In the minimal arc cover, we restore the direction of the arcs and denote the resulting bipartite digraph $\widehat{G}$. It should also be noted that there are two types of star-like sub graphs. In the *star-like sub graph* of the first type, the root is contained in the first lobe V_1 (on the left in Figure 1) and in the star-like sub graph of the second type, the root is contained in the second lobe V_2 (on the right in Figure 1).

The main result of the work is an algorithm for constructing a smallest set of good arcs that turn $\widehat{G}$ into a strongly connected digraph. It is proved that the number of good arcs in a smallest set $p(\widehat{G}) = \max(|V_1|, |V_2|)$, where $|V_i|$ the number of vertices in the set V_i, $i = 1, 2$. This equation applies to a bipartite graph G with lobes V_1, V_2 and to acyclic digraph $\mathcal{G}$. This is due to the fact that "good arcs," turning the bipartite digraph $\widehat{G}$ into strongly connected, also turn the bipartite digraph G and the original acyclic digraph $\mathcal{G}$ into strongly connected.

Main results. Consider a bipartite digraph $\widehat{G}$, consisting of the set of unrelated M stars $G_1^1, \ldots, G_1^M$ with the root in the first lobe and N stars $G_2^1, \ldots, G_2^N$ with the root in the second lobe. Let m the number of leaves in the stars $G_1^1, \ldots, G_1^M$ and n the number of leaves in the stars $G_2^1, \ldots, G_2^N$. Then, it performs equality $|V_1| = M + n$, $|V_2| = m + N$. Figure 1 shows an example of a digraph $\widehat{G}$, consisting of stars G_1^1, G_1^2, G_2^1, G_2^2 in the case of $m = n = 6$, $M = N = 2$. Here, the upper vertices make up the first lobe, and the lower ones make up the second lobe.

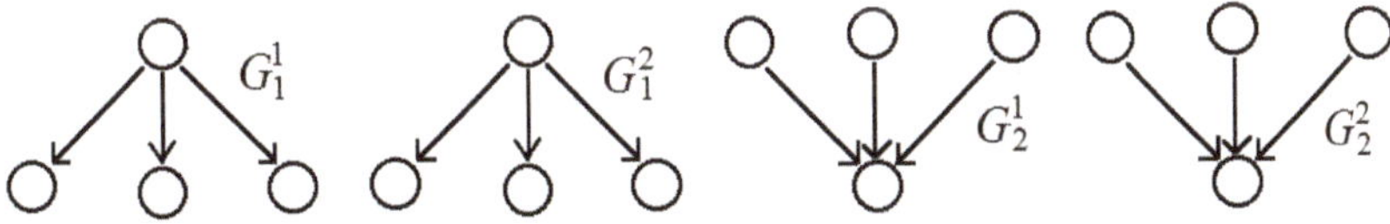

Figure 1. Unrelated stars G_1^1, G_1^2, G_2^1, G_2^2, $M = N = 2$, $m = n = 6$.

Theorem 1. *Equality* $|p(\widehat{G})| = \max(|V_1|, |V_2|)$ *is true.*

Proof of Theorem 1. When converting the digraph $\widehat{G}$, the number of good arcs, entering the roots of stars $G_1^1, \ldots, G_1^M$, must be no less than M, and entering the leaves of stars $G_2^1, \ldots, G_2^N$—no less than n. The number of good arcs, leaving the leaves of stars $G_1^1, \ldots, G_1^M$, must be no less than m, and leaving the roots of the stars $G_2^1, \ldots, G_2^N$ are no less than N. Therefore, the number of good arcs entering the vertices of the first lobe is no less than $M + n$, and leaving the vertices of the second lobe is no less than $m + N$. Additional incoming and outgoing arcs may coincide. Therefore, the minimum number of additional arcs is $|p(\widehat{G})| \geq \max(m + N, n + M)$. Now, we prove that $|p(\widehat{G})| = \max(M + n, m + N) = \max(|V_1|, |V_2|)$.

Let us first consider the case when the digraph $\widehat{G}$ consists only of stars $G_1^1, \ldots, G_1^M$ or only of stars $G_2^1, \ldots, G_2^N$. Let us add stars $G_1^1, \ldots, G_1^M$ by good arcs.

In the stars $G_1^1, \ldots, G_1^M$ with good arcs, we indicate the Hamiltonian cycle. It starts at the root of the star G_1^1, passes sequentially through all the leaves of this star, goes to the root of the star G_1^2, etc., from the last leaf of the star G_1^M to the root of the star G_1^1. As a result, we transform the stars $G_1^1, \ldots, G_1^M$ into a strongly connected digraph with the number of additional arcs $m = \max(m + 0, 0 + M)$ (Figure 2). If $M = 1$, then the last leaf of the star

G_1^1 is connected by the good arc with its root.

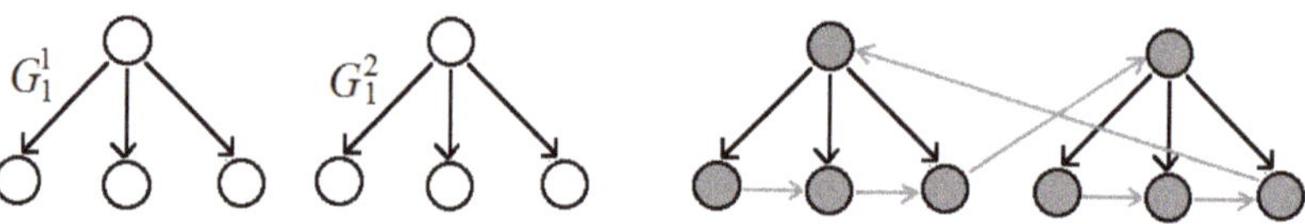

Figure 2. A strongly connected digraph constructed from the stars G_1^1, G_1^2 (right).

Similarly, in the digraph $\widehat{G}$, consisting only of stars $G_2^1, \ldots, G_2^N$, let us add these stars with good arcs sequentially connected by their leaves. In addition, from the root of the star G_2^k, we will draw a good arc to the first leaf of the star G_2^{k+1}, $k = 1, \ldots, N-1$, and from the root of the star G_2^N to the first leaf of the star G_2^1. If $N = 1$, we connect the root of the star G_2^1 with its first leaf.

In the stars $G_2^1, \ldots, G_2^N$ with additional arcs, we indicate the Hamiltonian cycle. It starts in the first leaf of the star G_2^1, passes sequentially through all its leaves, and goes to its root, then goes to the first leaf of the star G_2^2, etc. From the root of the star G_2^N, the path continues to the first leaf of the star G_2^1. As a result, we transform the stars $G_2^1, \ldots, G_2^N$ into a strongly connected digraph with the number of additional arcs $n = \max(0 + N, n + 0)$ (Figure 3). If $N = 1$, we connect the root of the star G_2^1 by the good arc with its first leaf.

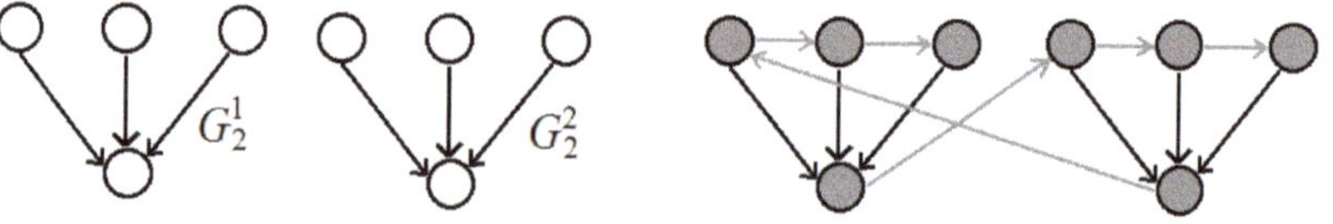

Figure 3. A strongly connected digraph constructed from the stars G_2^1, G_2^2 (right).

Let us now consider the case when $MN > 0$, i.e., in the digraph $\widehat{G}$, there are stars of both the first and the second types. Denote W_1 the set of all vertices in the stars $G_1^1, \ldots, G_1^M$ and W_2 the set of all vertices in the stars $G_2^1, \ldots, G_2^N$. Let us introduce a good arc, coming out of the root of the star G_2^N and entering some leaf of the star G_2^{N-1}, an arc coming out of the root of the star G_2^{N-1} and entering some leaf of the star G_2^{N-2}, etc., good arc, coming out of the root of the star G_2^1 and entering the root of the star G_1^M, good arc, coming out of some leaf of the star G_1^M and entering the root of the star G_1^{M-1}, coming out of any leaf of the star G_1^{M-1} and entering the root of the star G_1^{M-2}, etc., good arc, coming out of any leaf of the star G_1^2 and entering the root of the star G_1^1. Let us call the introduced good arcs and their incident vertices marked (see Figure 4, highlighted in grey). It is obvious that from any labelled, and hence from any vertex of the star of the set $G_2^1, \ldots, G_2^N$, there is a path to any vertex of the star of the set $G_1^1, \ldots, G_1^M$. We will denote this statement $W_2 \Rightarrow W_1$.

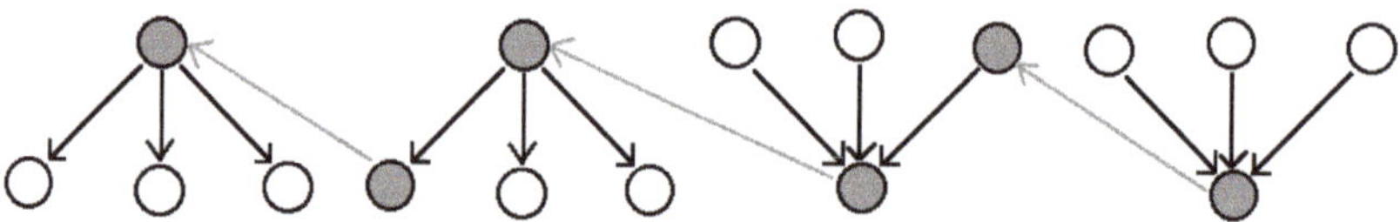

Figure 4. Introduction of $M + N - 1$ additional arcs.

The number of labelled arcs connected the vertices of the star $G_1^1, \ldots, G_1^M$, is $M - 1$, and the arcs connected the vertices of the star $G_2^1, \ldots, G_2^N$, is $N - 1$. Then, the total number of marked arcs taking into account the arc from the star G_2^1 to the star G_1^M is $M - 1 + N - 1 + 1 = M + N - 1$.

The total number of unlabelled vertices in the stars $G_1^1, \ldots, G_1^M$ is $m - (M - 1)$, the number of unlabelled vertices in the stars $G_2^1, \ldots, G_2^N$ is $n - (N - 1)$. From each unlabelled vertex of the set W_1 (see Figure 5), let us draw good arc to some unlabelled vertex of the set W_2 so that each unlabelled vertex of the set W_2 includes an arc from some vertex of the set W_1. Thus, the number of additionally introduced good arcs is equal to $\max(m - (M - 1), n - (N - 1))$. Therefore, the total number of good arcs becomes equal to $\max(m - (M - 1), n - (N - 1)) + M + N - 1 = \max(m + N, n + M)$.

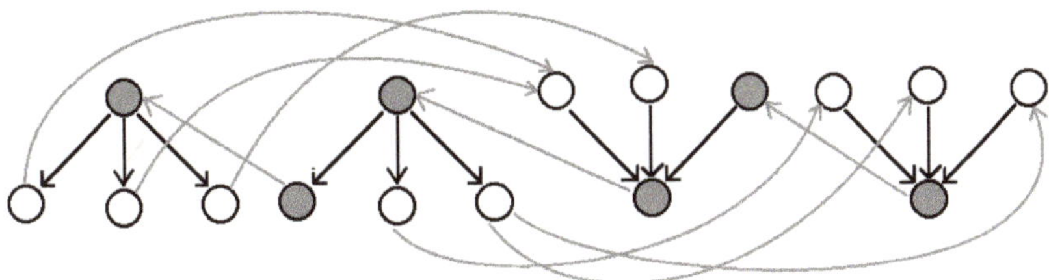

Figure 5. Introduction of $\max(m - (M - 1), n - (N - 1))$ good arcs.

We prove that the introduction of all good arcs into the stars $G_1^1, \ldots, G_1^M$, $G_2^1, \ldots, G_2^N$ transforms these stars into a strongly connected digraph. Let us take an arbitrary unlabelled vertices $v_1 \in W_1$, $v_2 \in W_2$ and draw the path through unlabelled vertices v_1, v_2', v_1', v_2, where $v_2' \in W_2$—the vertex connected with a vertex v_1 by good arc, and $v_1' \in W_1$—the vertex connecting with the vertex v_2 by good arc. Since from any labelled vertex of the set W_1 it is possible to draw a path to some unlabelled vertex of this set and from any unlabelled vertex of the set W_2 it is possible to draw good arc to some labelled vertex of this set, then it is possible to draw a path from any vertex of the set W_1 to any vertex of the set W_2, i.e., $W_1 \Rightarrow W_2$. Then, from the relations $W_1 \Rightarrow W_2$, $W_2 \Rightarrow W_1$ we get $W_1 \cup W_2 \Rightarrow W_1 \cup W_2$. Therefore, constructing from the stars $G_1^1, \ldots, G_1^M$, $G_2^1, \ldots, G_2^N$ digraph with the entered $\max(m + N, n + M)$ good arcs, is strongly connected. The statement of Theorem 1 is fully proved. $\square$

Theorem 2. *For a bipartite digraph G, the minimum number of good arcs, that turn it into a strongly connected digraph is determined by the equality*

$$|p(G)| = \max(|V_1|, |V_2|). \tag{1}$$

Proof of Theorem 2. From Theorem 1, the equalities

$$|V_1| = n + M, \quad |V_2| = m + N, \quad |p(\widehat{G})| = \max(|V_1|, |V_2|)$$

follow. From the definition of a minimal arc cover $\widehat{G}$, it follows that the set of its vertices coincides with the set of vertices in the bipartite digraph G. And the set of arcs in $\widehat{G}$ is contained in the set of arcs in G, therefore $\max(|V_1|, |V_2|) = |p(\widehat{G})| \geq |p(G)|$. However, since $|p(G)| \geq \max(|V_1|, |V_2|)$, the equality (1) is fulfilled. Theorem 2 is proved. $\square$

Remark 2. *Using the algorithm for proving Theorem 1, it is possible to construct a smallest set $p(G)$ of good arcs that transform a bipartite digraph G into a strongly connected digraph $\widetilde{G}$. Thus, a constructive solution is given to the problem of determining the smallest set of good arcs in a bipartite digraph G.*

Theorem 3. *For an acyclic digraph $\mathcal{G}$, the minimum number of good arcs that turn it into a strongly connected one is determined by the equality*

$$|p(\mathcal{G})| = \max(|V_1|, |V_2|).$$

Proof of Theorem 3. By arcs from the smallest set $p(G)$ of good arcs, we connect the vertices of the sets V_1, V_2 into $\mathcal{G}$. We obtain from the acyclic digraph $\mathcal{G}$ a strongly connected digraph in which the minimum number of good arcs $|p(\mathcal{G})| = \max(|V_1|, |V_2|)$. Theorem 3 is proved. $\square$

Remark 3. *Assume that the acyclic digraph $\mathcal{G}$ has an isolated vertex that no arcs enter into it or exit from it. Then, we may fictitiously introduce this vertex into the first and second lobes and connect these vertices with a fictitious arc. Then all further constructions are saved.*

4. Recurrent Algorithm for Class Allocation Cyclic Equivalence

This section provides one of the algorithms for converting a digraph into an acyclic digraph by allocating cyclic equivalence classes in it. There are different algorithms to solve this problem, see for example [12,28], etc. In this section, we show sequential algorithm in which at each step new vertex and arcs connecting it with previously introduced are added to the digraph. This algorithm was convenient to deal with protein networks in numerical examples [29,30].

Let us say that two vertices of a digraph are cyclically equivalent if they are included in any cycle contained in it. On the set of cyclic equivalence classes (clusters), a partial order relation is defined $v \succeq w$, if there is a path from the cluster v to the cluster w. We define a zero-one matrix $\|a(v,w)\|$ by the condition $a(v,w) = 1 \iff v \succeq w$. Then, the algorithm for determining the set of clusters and the matrix a, specifying the partial order $\succeq$ on it, is based on the following recurrent procedure [29].

Let all vertices in the original digraph be numbered: $1, 2, \ldots, n$. At step 1, a single cluster is constructed consisting of a vertex 1 and a partial order matrix a, consisting of a single element $a(1,1) = 1$. Suppose that at step $t - 1$, clusters and a matrix specifying a partial order $\succeq$ between them are given. Then, at the step t, the vertex t and the good arcs connecting this vertex to the already specified clusters are added. Then, in a digraph consisting of clusters constructed at step $t - 1$ and paths between them, after adding a vertex t and good arcs connecting it to already constructed clusters, sets of clusters B_1, B_2, B are determined (see Figure 6, left). The set B_1 contains clusters into which there is a path from the vertex t. Similarly, the set B_2 contains clusters from which there is a path to the vertex t. All other clusters fall into the set B, and from them there can be paths only to the clusters of the set B_1 and paths can exist in them only from clusters of the set B_2 (see Figure 6, left). Then, at step t, a new cluster $[t]$ is built, consisting of the vertex t and the clusters of the set $A = B_1 \cap B_2$, and the paths between the remaining clusters are shown in Figure 6, on the right. Then, the matrix a of partial order $\succeq$ is defined by Table 1. In this table, rectangular sub matrices 0 consist of only zeros, rectangular sub matrices 1 consist of only ones, and rectangular matrices denoted by values at step $t - 1$ repeat the corresponding sub matrices of the matrix a at step $t - 1$ (see [29]).

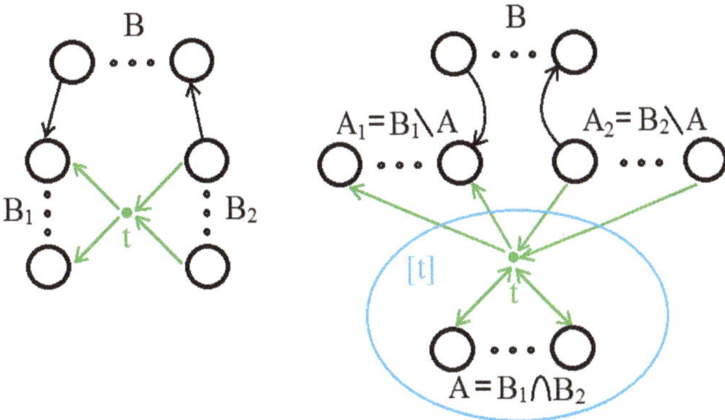

Figure 6. The algorithm of transition from step $t-1$ to step t for clusters.

Table 1. Algorithm of transition from step $t-1$ to step t for a matrix of partial order a.

Matrice Partial Order	Clusters Set A_1	Clusters Set $[t]$	Clusters Set A_2	Clusters Set B
clusters of set A_1	meanings on step $t-1$	0		0
clusters of set $[t]$		1		
clusters of set A_2			meanings on step $t-1$	
clusters of set B	meanings on step $t-1$	0		meanings on step $t-1$

As a result of such clustering, the original digraph is transformed into a digraph with a set of cluster vertices. An arc is drawn between two clusters if at least one arc exists between them in the original digraph.

5. Discussions

Thus, the tasks set in the paper are solved by reducing to the problem of the maximum flow and the minimum section. This allows us to use Ford–Fulkerson algorithms [4,5] and their modifications [22,23], which avoid the need to solve NP-problems. To do this, either the arc throughput is determined in a special way, or good arcs are introduced not for the entire acyclic digraph, but for its arcs coverage, which greatly simplifies the task. We also note that the optimization tasks considered in the paper do not always have a single solution. However, the proposed algorithms allow us to obtain some of these solutions.

Author Contributions: Conceptualization, G.T. and V.B.; methodology and formal analysis, G.T.; data curation and writing, V.B.; All authors have read and agreed to the published version of the manuscript.

Funding: This research received no external funding.

Institutional Review Board Statement: Not applicable.

Informed Consent Statement: Not applicable.

Data Availability Statement: Data supporting reported results were obtained by Victor Bulgakov.

Conflicts of Interest: The authors declare no conflict of interest.

References

1. Harary, F.; Norman, R.Z.; Cartwright, D. *Structural Models: An Introduction to the Theory of Directed Graphs*; Wiley: New York, NY, USA, 1965.
2. Wasserman, S.S. Models for Binary Directed Graphs and Their Applications. *Adv. Appl. Prob.* **1978**, *10*, 803–818. [CrossRef]
3. Bang-Jensen, J.; Gutin, G. *Classes of Directed Graphs*; Springer: Berlin/Heidelberg, Germany, 2018.
4. Ford, L.R., Jr.; Fulkerson, D.R. Maximal Flow Through a Network. *Can. J. Math.* **1956**, *8*, 399–404. [CrossRef]
5. Ford, L.R., Jr.; Fulkerson, D.R. A Simple Algorithm for Finding Maximal Network Flows and an Application to the Hitchcock Problem. *Can. J. Math.* **1957**, *9*, 210–218. [CrossRef]
6. Cormen, T.H.; Leiserson, C.E.; Rivest, R.L.; Stein, C. *Introduction to Algorithms*, 3rd ed.; The MIT Press: New York, NY, USA, 2009.
7. Cook, W.J.; Cunningham, W.H.; Pulleyblank, W.R.; Schriver, A. Combinatorial Optimization. *Wiley Ser. Discret. Math. Optim.* **2011**, *33*, 48–49.
8. Potapov, A.P.; Goemann1, B.; Wingender1, E. The pairwise disconnectivity index as a new metric for the topological analysis of regulatory networks. *BMC Bioinform.* **2008**, *9*, 227. [CrossRef] [PubMed]
9. Sheridan, P.; Kamimura, T.; Shimodaira, H. A Scale-Free Structure Prior for Graphical Models with Applications in Functional Genomics. *PLoS ONE* **2010**, *5*, e13580. [CrossRef]
10. Georgiadis, L.; Giuseppe, F.I.; Parotsidis, N. Strong Connectivity in Directed Graphs under Failures, with Applications. *SIAM J. Comput.* **2020**, *49*, 865–926. [CrossRef]
11. Melo, B.; Brandao, I.; Tomei, C.; Guerreiro, T. Directed graphs and interferometry. *J. Opt. Soc. Am. B* **2020**, *37*, 2199–2208. [CrossRef]
12. Mezic, I.; Fonoberov, V.A.; Fonoberova, M.; Sahai, T. Spectral Complexity of Directed Graphs and Application to Structural Decomposition. *Complexity* **2019**, *9610826*. [CrossRef]
13. Marques, A.G.; Segarra, S.; Mateos, G. Signal Processing on Directed Graphs: The Role of Edge Directionality When Processing and Learning From Network Data. *IEEE Signal Process. Mag.* **2020**, *37*, 99–116. [CrossRef]
14. Dokka, T.; Kouvela, A.; Spieksma, F.C.R. Approximating the multi-level bottleneck assignment problem. *Oper. Res. Lett.* **2012**, *40*, 282–286. [CrossRef]
15. Zwick, U. The smallest networks on which the Ford-Fulkerson maximum flow procedure may fail to terminate. *Theor. Comput. Sci.* **1995**, *148*, 165–170. [CrossRef]
16. Backman, S.; Huynh, T. Transfinite Ford–Fulkerson on a finite network. *Computability* **2018**, *7*, 341–347. [CrossRef]
17. Allen, B.; Sample, C.; Jencks, R.; Withers, J.; Steinhagen, P.; Brizuela, L.; Kolodny, J.; Parke, D.; Lippner, G.; Dementieva, Y.A. Transient amplifiers of selection and reducers of fixation for death-Birth updating on graphs. *PLoS Comput. Biol.* **2020**, *16*, e1007529. [CrossRef]
18. Bulgakov, V.P.; Avramenko, T.V.; Tsitsiashvili, G.S. Critical analysis of protein signaling networks involved in the regulation of plant secondary metabolism: Focus on anthocyanins. *Crit. Rev. Biotechnol.* **2017**, *37*, 685–700. [CrossRef]
19. Bulgakov, V.P.; Wu, H.C.; Jinn, T.L. Coordination of ABA and Chaperone Signalling in Plant Stress Responses. *Trends Plant Sci.* **2019**, *24*, 636–651. [CrossRef]
20. Fàbregas, N.; Lozano-Elena, F.; Blasco-Escámez, D.; Tohge, T.; Martínez-Andújar, C.; Albacete, A.; Osorio, S.; Bustamante, M.; Riechmann, J.L.; Nomura, T.; et al. Overexpression of the vascular brassinosteroid receptor BRL3 confers drought resistance without penalizing plant growth. *Nat. Commun.* **2018**, *9*, 4680. [CrossRef]
21. Bulgakov, V.P.; Avramenko, T.V. Linking Brassinosteroid and ABA Signaling in the Context of Stress Acclimation. *Int. J. Mol. Sci.* **2020**, *21*, 5108. [CrossRef] [PubMed]
22. Edmonds, J.; Karp, R.M. Theoretical Improvements in Algorithmic Efficiency for Network Flow Problems. *J. ACM* **1972**, *19*, 248–264. [CrossRef]
23. Arlazarov, V.L.; Dinitz, E.A.; Ilyashenko, Y.S.; Karzanov, A.V.; Karpenko, S.M.; Kirillov, A.A.; Konstantinov, N.N.; Kronrod, M.A.; Kuznetsov, O.P.; Okun', L.B.; et al. Georgy Maksimovich Adelson-Velsky (obituary). *Russ. Math. Surv.* **2014**, *69*, 743–751. [CrossRef]
24. Monjurul Alom, B.M.; Someresh, D.; Islam, M.S. Finding the Maximum Matching in a Bipartite Graph. *DUET J.* **2010**, *1*, 33–36.
25. Tutte, W.T. The method of alternating paths. *Combinatorica* **1982**, *2*, 325–332. [CrossRef]
26. Pemmaraju, S.; Skiena, S. *Computational Discrete Mathematics: Combinatorics and Graph Theory with Mathematica*; Cambridge University Press: Cambridge, UK, 2003.
27. Tsitsiashvili, G.; Osipova, M.A. Cluster Formation in an Acyclic Digraph Adding New Edges. *Reliab. Theory Appl.* **2021**, *16*, 37–40.
28. Tarjan, R. Depth-first Search and Linear Graph Algorithms. *SIAM J. Comput.* **1972**, *1*, 146–160. [CrossRef]
29. Tsitsiashvili, G. Sequential algorithms of graph nodes factorization. *Reliab. Theory Appl.* **2013**, *8*, 30–33.
30. Tsitsiashvili, G.; Kharchenko, Y.N.; Losev, A.S.; Osipova, M.A. Analysis of Hubs Loads in Biological Networks. *Reliab. Theory Appl.* **2014**, *9*, 7–10.

 *mathematics*

Article

On Reliability Function of a k-out-of-n System with Decreasing Residual Lifetime of Surviving Components after Their Failures

Vladimir Rykov [1,2,3], Nika Ivanova [1,4,*], Dmitry Kozyrev [1,4] and Tatyana Milovanova [1]

[1] Department of Applied Probability and Informatics, Peoples' Friendship University of Russia (RUDN University), 6 Miklukho-Maklaya Str., 117198 Moscow, Russia
[2] Department of Applied Mathematics and Computer Modelling, National University of Oil and Gas "Gubkin University", 65 Leninsky Prospect, 119991 Moscow, Russia
[3] Institute for Transmission Information Problems (Named after A.A. Kharkevich) RAS, Bolshoy Karetny, 19, GSP-4, 127051 Moscow, Russia
[4] V.A. Trapeznikov Institute of Control Sciences of Russian Academy of Sciences, 65 Profsoyuznaya Str., 117997 Moscow, Russia
* Correspondence: nm_ivanova@bk.ru

Abstract: We consider the reliability function of a k-out-of-n system under conditions that failures of its components lead to an increase in the load on the remaining ones and, consequently, to a change in their residual lifetimes. Development of models able to take into account that failures of a system's components lead to a decrease in the residual lifetime of the surviving ones is of crucial significance in the system reliability enhancement tasks. This paper proposes a novel approach based on the application of order statistics of the system's components lifetime to model this situation. An algorithm for calculation of the system reliability function and two moments of its uptime has been developed. Numerical study includes sensitivity analysis for special cases of the considered model based on two real-world systems. The results obtained show the sensitivity of system's reliability characteristics to the shape of lifetime distribution, as well as to the value of its coefficient of variation at a fixed mean.

Keywords: k-out-of-n system; dependent failures; order statistics; reliability characteristics; sensitivity analysis

MSC: 60H99

Citation: Rykov, V.; Ivanova, N.; Kozyrev, D.; Milovanova, T. On Reliability Function of a k-out-of-n System with Decreasing Residual Lifetime of Surviving Components after Their Failures. *Mathematics* **2022**, *10*, 4243. https://doi.org/10.3390/math10224243

Academic Editors: Gurami Tsitsiashvili and Alexander Bochkov

Received: 25 October 2022
Accepted: 11 November 2022
Published: 13 November 2022

1. Introduction and Motivation

Ensuring the reliability of systems, objects, and processes is one of the main goals in their creation and further operation. Redundancy serves this aim, and a k-out-of-n : F model is a very popular configuration for it. This is a model of a system that consists of n components in parallel that fails when at least k of them fail. Hereinafter, we will use this notation omitting the symbol "F".

Due to the wide range of practical applications of k-out-of-n systems, many papers have been devoted to their study. The bibliography on the related topics is extensive (see Trivedi [1], Chakravarthy et al. [2] and the bibliography therein). For a brief overview of further investigations, see, for example, [3] by Rykov et al. An overview of recent publications on k-out-of-n multi-state systems can be found in [4]. Furthermore, the k-out-of-n systems with several types of failure have been considered in [5,6]. In the 1980s in [7,8] for the investigation of heterogeneous systems, Ushakov proposed the method of Universal Generating functions. At present, it has become a very popular technique and has been used in different applications (see, for example, a monograph by Levitin [9] and the bibliography therein). Recently, in [10], Kala proposed new sensitivity measures for the system's reliability function based on the entropy of its structural function. Engineering applications of this model to the study of real-world systems can be found in [11] for the reliability study of some structures in the oil and gas industry. In [12], the model is used for reliability analysis of a remote monitoring system of underwater sections of gas pipeline, and in [13] for the reliability study of a rotary-wing flight module of a high-altitude telecommunications platform.

Another interesting line of research within the framework of the problem of system reliability enhancement is the prediction of the remaining useful life (RUL), which is an indispensable indicator to measure the degradation process of system components. In [14], a novel adaptive approach based on Kalman filter and expectation maximum with Rauch–Tung–Striebel was proposed to solve the problem of the RUL prediction of lithium-ion battery which is critical for the normal operation of electric vehicles [15].

A data-driven RUL prediction approach based on deep learning was proposed in [16] and verified by two real-world datasets—the aircraft engines dataset and the actual milling machine dataset.

Recently, an interesting approach of stress–strength reliability characteristic study was proposed (see, especially, [17–19]). It is interesting to study this index with respect to its sensitivity to both stress and strength distribution. We do not touch this approach here, but it will be in our plans in future.

In paper [13], a wide range of issues was posed for the study of systems whose failure depend not only on the number of failed components, but also on their location in the system. Moreover, it is also very important to take into account that failures of system components lead to the increase in the load on the remaining ones. A simple load-sharing model, in which the lifetime is exponentially distributed and the load from the failed components is distributed proportionally among the survivors, is considered in [20] through the example of a 2-out-of-3 system. A load-sharing k-out-of-n : G system with identical components and arbitrary distribution of lifetime under the equal load-sharing rule in the context of semi-Markov embedded processes was studied in [21].

The study of a k-out-of-n model in which failed components do not affect the residual lifetime of surviving components, using order statistics, is considered in [22]. On the other hand, the increase in the load on working components after the stop of functioning of the failed ones can lead to the decrease in their residual lifetime. Such a problem has been studied in our previous papers [12,23]. In addition, in [24] this problem was modeled by the changing in components' hazard rate function.

The application of order statistics to the study of k-out-of-n models is not new [1,25]. Previously, in [26], the so-called sequential order statistics (which is some extension of ordinary order statistics) were considered for the study of a k-out-of-n system, in which a failure of any component can affect other components, so that their basic failure rate is corrected in relation to the number of previous failures. A similar model of the impact of a component's failure on the functioning of the survived ones has been developed for example in [27,28], where it was supposed that the failure of any component influences the others, so that their failure rate is adjusted with respect to the number of preceding failures.

However, the problem of system failure, associated with a change in the residual component lifetime, depending on the increase in load after the failure of any component, has not yet been solved. Thus, the present article is devoted to the solution of this problem. The novelty of this investigation consists of the following:

- we perform the reliability study of a k-out-of-n system, whose component failures change residual lifetime of the other components;
- in the current paper, despite the fact that order statistics have already been applied to the study of k-out-of-n system reliability characteristics, we propose a novel application of order statistics to study of the lifetimes of components and the whole system.

The paper is organized as follows. In the next section, the problem is set up, the main notations and some practical examples of k-out-of-n models are given. Then, in Section 3 the necessary preliminaries are introduced and in Section 4 the general procedure for the solution of the stated problem is proposed. The numerical study of different scenarios for the investigation of a 2-out-of-6 system is made in Section 5. In conclusion, directions for further research are outlined.

2. State of the Problem: Notations and Examples

2.1. Problem Setting

Usually, real-world redundant systems are constructed based on the same type of components. Thus, we consider a k-out-of-n system that consists of n identical components in parallel and fails if at least k of them fail. At that point, it is supposed that the failure of an i-th component for $i < k$ leads to the increase of load on the others and therefore to the decrease of their residual lifetimes. It is modeled by multiplying the residual lifetime of the surviving components by some weighting factor $c_i < 1$, $(i = \overline{1, k-1})$. We will consider the system operation up to its first failure.

In the present paper, the main reliability characteristics of such a system are studied, namely:

– time T to the first failure of the system,
– reliability function $R(t) = \mathbf{P}\{T > t\}$ of the system,
– its two first moments,
– high confidence quantiles;
– sensitivity analysis of the system's reliability function to the shapes of its components' lifetime distribution.

2.2. Notations: Assumptions

To study the system, introduce the following notations:

• $\mathbf{P}\{\cdot\}$, $\mathbf{E}[\cdot]$ are symbols of probability and expectation;
• $A_i : (i = \overline{1, n})$ is the series of components' lifetimes, which are supposed to be independent identically distributed (iid) random variables (rv);
• $A(t) = \mathbf{P}\{A_i \le t\}$ is their common cumulative distribution function (cdf);
• j is the system state, which means the number of failed components;
• $E = \{j = \{0, 1, \ldots, k\}\}$ is the set of the system states.

Under the set of states E, define a stochastic process $J = \{J(t) : t \ge 0\}$ by the expression

$$J(t) = j, \ \text{if in time } t \text{ the system is in the state } j \in E$$

and denote by T and $R(t)$ time to the first system failure and the reliability function, respectively,

$$T = \inf\{t : J(t) = k\}, \quad R(t) = \mathbf{P}\{T > t\}.$$

2.3. Examples

As mentioned in the Introduction, k-out-of-n models have a wide sphere of applications (see [1] and others), including the study of energy (see [11,12]), and telecommunication [13] problems. Let us focus on two examples of applying the k-out-of-n model. In the numerical analysis, we will use these examples for the special case of $n = 6$, $k = 2$.

2.3.1. A Flight Module of a Tethered High-Altitude Telecommunication Platform

As an application example of the proposed k-out-of-n model, consider the model of a multi-copter flight module, which is part of the tethered high-altitude telecommunications platform [13]. The main area of its application is solving problems related to the long-term operation (tens of hours) without lowering the unmanned flight module to the ground. Therefore, unlike autonomous Unmanned Aerial Vehicles (UAVs) reliability parameters are of crucial importance for the tethered UAV-based high-altitude platforms.

A multi-rotor UAV is a system consisting of n rotors arranged uniformly in a circle and pairwise symmetrically with respect to the center of the circle [29]. The multi-copter may malfunction due to the failure of the propeller engines. There are various modifications of multi-rotor UAVs. The most common architectures are quad-, hexa-, and octocopters.

The higher the redundancy ratio, the higher the reliability of the system. , Therefore, in practice, flight modules with 6 or 8 rotors are most often used. In this example, we consider a hexacopter as a hot standby system consisting of $n = 6$ components (rotors) that work and fail independently of each other (see Figure 1).

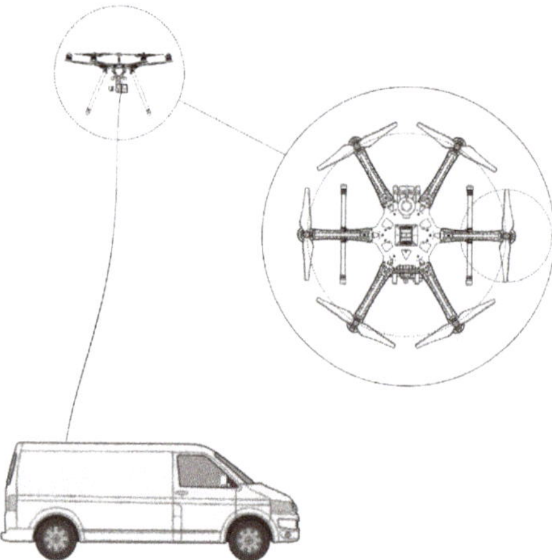

Figure 1. An unmanned hexacopter flight module of a tethered high-altitude telecommunications platform.

If the location of the failing components is not taken into account, this system fails when $k = 2$ out of 6 rotors fail.For practical use, various reliability characteristics of such a system, including those considered in the general model, are of interest.

2.3.2. An Automated System for Remote Monitoring of a Sub-Sea Pipeline

As another application example of the k-out-of-n model, we consider an automated system for remote monitoring of a sub-sea pipeline. This system has been considered in [12], where its description has been given in details. One of the main parts of this system is an Unmanned Underwater Vehicle (UUV), the structure of which is illustrated in Figure 2.

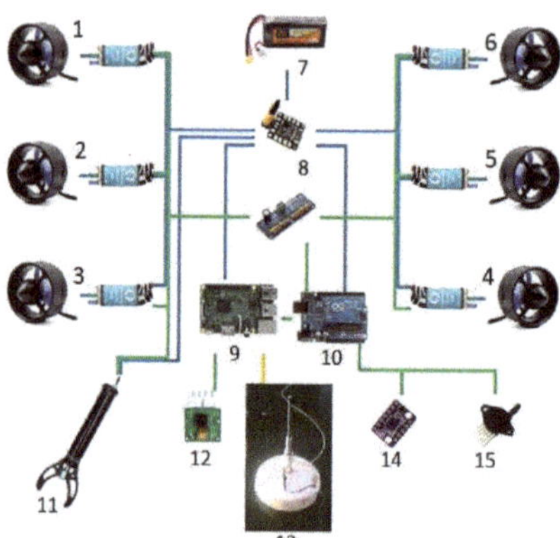

Figure 2. An unmanned multi-functional underwater vehicle.

The UUV consist of 6 motors, indicated by numbers 1–6, which allow it to rise, fall and move in various directions, including along the pipeline. The UUV is equipped with various devices, indicated by numbers 7–15, for receiving and transmitting information about the state of the pipe. In paper [12] the reliability function of this model has been studied in two scenarios:

(1) in the case, when the system's failure depends only on the number of its failed components. At that point, it is assumed that the device can perform its functions until at least 3 of its engines fail;

(2) in the case, when the system's failure depends also on the position of the failed components in the system. At that point, the UUV can perform its functions as long as at least two engines located on opposite sides, or any three engines are operational. Therefore, it could be considered to be a combination of $3 + 1$-out-of-6 : F and 5-out-of-6 : F systems. For such a system, the special notation such as $(5, 3 + 1)$-out-of-6 : F system was used.

However, the influence of the number of failed components on the residual lifetimes of the survived ones was not taken into account earlier. In the current paper, this model has been studied under the condition that failed components reduce the residual lifetime of surviving system's components.

3. Distribution of the System's Time to Failure

3.1. Preliminaries

It is evident that if a k-out-of-n system's failure depends only on several its failed components, it coincides with the k-th order statistic from n iid rv A_i $(i = \overline{1, n})$ with a given cdf $A(t)$. For simplicity, further we will denote order statistics $A_{(1)} \leq \cdots \leq A_{(k)} \leq \cdots \leq A_{(n)}$ of iid rv A_i $(i = \overline{1, n})$ by X_i, i.e., $X_i = A_{(i)}$ and $X_1 \leq \cdots \leq X_k \leq \cdots \leq X_n$. Distributions of order statistics are well studied (see, for example, [30]), where it was shown that the joint probability density function (pdf) $f_n(x_1, \ldots x_n)$ of all order statistics $X_1 \leq X_2 \leq \cdots \leq X_n$ from n iid rv $A_1, A_2, \ldots, A_n$ with a given pdf $a(x)$ has the following form:

$$f_n(x_1, x_2, \ldots, x_n) = n! a(x_1) a(x_2) \cdots a(x_n) \quad (x_1 \leq x_2 \leq \cdots \leq x_n).$$

By integration of this pdf with respect to last $n - k$ variables one can simply find the joint pdf $f_k(x_1, \ldots x_k)$ of the first k order statistics $X_1 \leq X_2 \leq \cdots \leq X_k$ from the n iid rv A_i $(i = \overline{1, n})$ in the domain $x_1 \leq x_2 \leq \ldots \leq x_k$ in the form

$$f_k(x_1, x_2, \ldots, x_k) = \frac{n!}{(n-k)!} a(x_1) a(x_2) \ldots a(x_k)(1 - A(x_k))^{n-k}. \tag{1}$$

However, if a failure of one of the system's components leads to the change in the residual lifetimes of all survived components, then their distributions are also changed.

3.2. Transformation of Order Statistics

Following the proposed model of the influence of components' failures on the residual lifetime of survivors, they are reduced by multiplying by some constant c_i depending on the number of failed components. Denote by Y_i $(i = \overline{1, k})$ the time of an i-th component failure under the conditions of increasing the load on survived components. To simplify the representation of these values in terms of order statistics $X_1 \leq X_2 \leq \cdots \leq X_n$, we introduce the following notations,

$$C_1 = (1 - c_1), \quad C_2 = c_1(1 - c_2), \quad \ldots, \quad C_{k-1} = c_1 \cdots c_{k-2}(1 - c_{k-1}), \quad C_k = c_1 \cdots c_{k-1}.$$

In these notations, the following theorem holds.

Theorem 1. *The time to the considered system failure Y_k is a linear function of order statistics of the following form:*

$$Y_k = C_1 X_1 + C_2 X_2 + \cdots + C_{k-1} X_{k-1} + C_k X_k. \tag{2}$$

Proof. To calculate the time of the system failure, we slightly expand the problem statement and calculate the successive time moments Y_i $(i = \overline{1,k})$ of failures of the system's components under conditions of increasing load on the surviving components. To do that, we use a recursive procedure and denote by $X_i^{(j)}$ the expected time moment of the i-th failure after the failure of the j-th component $(i > j)$.

Thus, to start the induction, we have $X_i^{(0)} = X_i$. After the first failure of a component at time $Y_1 = X_1^{(0)} = X_1$ all residual lifetimes of surviving components that equal $X_i - X_1$ for $i > 1$ decrease by a factor of c_1, and therefore the expected failure times $X_i^{(1)}$ for $i > 1$ take the form

$$X_i^{(1)} \;=\; X_1^{(0)} + c_1(X_i^{(0)} - X_1^{(0)}) = (1 - c_1)X_1 + c_1 X_i, \quad i = \overline{2,k}.$$

Therefore $Y_2 = X_2^{(1)} = (1 - c_1)X_1 + c_1 X_2$.

Similarly, after the j-th failure at time $Y_j = X_j^{(j-1)}$, the residual lifetimes $X_i^{(j-1)} - X_j^{(j-1)}$ of all surviving components for all $i > j$ decrease by a factor c_j, $0 < c_j < 1$ $(j = \overline{1,k})$ and the expected failure times of components take the following form:

$$X_i^{(j)} \;=\; X_i^{(j-1)}, \; \forall i \le j,$$
$$X_i^{(j)} \;=\; X_j^{(j-1)} + c_j(X_i^{(j-1)} - X_j^{(j-1)}) = (1 - c_j)X_j^{(j-1)} + c_j X_i^{(j-1)}, \; \forall i > j.$$

Thus, the expected failure times of the system components Y_j $(j = \overline{1,k})$ under conditions of load redistribution equal to $Y_j = X_j^{(j-1)}$ $(j = \overline{1,k})$. Expressing $X_i^{(j)}$ in terms of the original order statistics, we obtain the following expression for $i > j$:

$$
\begin{aligned}
X_i^{(j)} &= (1-c_1)X_1 + c_1(1-c_2)X_2 + c_1 c_2(1-c_3)X_3 + \cdots + c_1\cdots c_{j-1}(1-c_j)X_j + c_1 \cdots c_j X_i \\
&= \sum_{l=1}^{j} c_1 \cdots c_{l-1}(1-c_l)X_l + c_1 \cdots c_j X_i.
\end{aligned}
\tag{3}
$$

Supposing that the last expression is true for a given j check it for all $i > j$:

$$
\begin{aligned}
X_i^{(j+1)} &= (1-c_1)X_1 + c_1(1-c_2)X_2 + c_1 c_2(1-c_3)X_3 + \cdots + c_1 \cdot \ldots \cdot c_{j-1}(1-c_j)X_j \\
&+ c_1 \cdot \ldots \cdot c_j(1-c_{j+1})X_{j+1} + c_1 \cdot \ldots \cdot c_j c_{j+1} X_i = \\
&= \sum_{l=1}^{j} c_1 \cdots c_{l-1}(1-c_l)X_l + c_1 \cdots c_j(1-c_{j+1})X_{j+1} + c_1 \cdots c_{j+1} X_i = \\
&= \sum_{l=1}^{j+1} c_1 \cdots c_{l-1}(1-c_l)X_l + c_1 \cdots c_{j+1} X_i.
\end{aligned}
\tag{4}
$$

Hence, by the principle of mathematical induction, the equality (3) holds for any j. In terms of the original order statistics X_i $(i = \overline{1,k})$, we obtain for all $j = \overline{1,k}$:

$$Y_j = X_j^{(j-1)} = (1-c_1)X_1 + c_1(1-c_2)X_2 + \cdots + c_1 \cdots c_{j-2}(1-c_{j-1})X_{j-1} + c_1 \cdots c_{j-1}X_j,$$

which, using the notation introduced earlier, leads to (2) for $j = k$, which completes the proof. $\square$

3.3. Distribution of the System Failure Time

Now move on to the calculation of the cdf $F_{Y_k}(y)$ of the system's time to failure Y_k under the condition of redistribution of the load on the components. We will do that by taking into account expression (2) for the time of the system failure in terms of order statistics X_i and using Formula (1) for the joint distribution of the first k order statistics.

To simplify the representation of this cdf we introduce the following notations,

$$
\begin{aligned}
z_0 &= y, \\
z_i &= z_i(y; x_1, \ldots, x_i) = \frac{y - C_1 x_1 + C_2 x_2 - \ldots - C_i x_i}{C_{i+1}} \quad (i = \overline{1, k-1}).
\end{aligned}
\tag{5}
$$

With these notations the following theorem holds.

Theorem 2. *The distribution of the system's time to failure for $y > 0$ is*

$$
\begin{aligned}
F_{Y_k}(y) &= \mathbf{P}\{Y_k < y\} \\
&= \frac{n!}{(n-k)!} \int_0^{z_0} a(x_1) dx_1 \int_{x_1}^{z_1} a(x_2) dx_2 \cdots \int_{x_{k-1}}^{z_{k-1}} a(x_k)(1 - A(x_k))^{n-k} dx_k.
\end{aligned}
\tag{6}
$$

Proof. According to Theorem 1 (see Formula (2)) the time Y_k of the system failure is the linear function of the first k order statistics

$$
Y_k = C_1 X_1 + C_2 X_2 + \cdots + C_{k-1} X_{k-1} + C_k X_k.
$$

Therefore, for cdf $F_{Y_k}(y)$ of rv Y_k in terms of pdf $f_k(x_1, \ldots, x_k)$ of the first k order statistics we obtain

$$
\begin{aligned}
F_{Y_k}(y) &= \mathbf{P}\{Y_k < y\} \\
&= \mathbf{P}\{C_1 X_1 + C_2 X_2 + \cdots + C_{k-1} X_{k-1} + C_k X_k < y\} \\
&= \int \cdots \int_{D(x_1, \ldots, x_k; y)} f_k(x_1, x_2, \ldots, x_k) dx_1 \ldots dx_k,
\end{aligned}
\tag{7}
$$

where the integration domain is

$$
D(x_1, \ldots, x_k; y) = \{0 \le x_1 \le \cdots \le x_k, \ C_1 x_1 + C_2 x_2 + C_3 x_3 + \ldots + C_{k-1} x_{k-1} + C_k x_k \le y\}.
$$

Let us represent this multidimensional integral as an iterated one. Taking into account that $x_1 \le x_2 \le \cdots \le x_k$, the integration domain can be transformed in the following way. For the last variable x_k from the inequality

$$
C_1 x_1 + C_2 x_2 + C_3 x_3 + \ldots + C_{k-1} x_{k-1} + C_k x_k \le y,
$$

it follows that

$$
x_k \le \frac{y - C_1 x_1 - C_2 x_2 - C_3 x_3 - \ldots - C_{k-1} x_{k-1}}{C_1 \cdots C_{k-1}} = z_{k-1}(y; x_1 \ldots x_{k-1}).
$$

Furthermore, taking into account that $x_{k-1} \le x_k$, from the last inequality, it follows that

$$
x_{k-1} \le x_k \le \frac{y - C_1 x_1 - C_2 x_2 - C_3 x_3 - \ldots - C_{k-1} x_{k-1}}{C_1 \cdots C_{k-1}}.
$$

From this inequality with the simple algebra one can find

$$
x_{k-1} \le \frac{y - C_1 x_1 - C_2 x_2 - \ldots - C_{k-2} x_{k-2}}{C_1 \cdots C_{k-2}} = z_{k-2}(y; x_1 \ldots, x_{k-2}).
$$

Following in the same way we obtain for variable x_2 the inequality

$$
\begin{aligned}
y &\geq (1-c_1)x_1 + c_1(1-c_2)x_2 + c_1c_2x_3 \geq \\
&\geq (1-c_1)x_1 + c_1(1-c_2)x_2 + c_1c_2x_2 = (1-c_1)x_1 + c_1x_2,
\end{aligned}
$$

from which it follows that

$$
x_2 \leq \frac{y - (1-c_1)x_1}{c_1},
$$

and, at last,

$$
y \geq (1-c_1)x_1 + c_1x_1 = x_1.
$$

It means that $0 \leq x_1 \leq y$. This argumentation shows that the integration domain $D(x_1, \ldots, x_k; y)$ in terms of notations (5) can be represented as

$$
D(x_1, \ldots, x_k; y) = \{x_{i-1} \leq x_i \leq z_i(y; x_1, \ldots x_{i-1}) \ (i = \overline{1,k})\}.
$$

Thus, using formula (1) for pdf $f_k(x_1, \ldots, x_k)$ for the first k order statistics and the above form of the integration domain, we can rewrite integral (7) for $y \geq 0$ as

$$
F_{Y_k}(y) = \frac{n!}{(n-k)!} \int_0^y a(x_1)dx_1 \int_{x_1}^{z_1} a(x_2)dx_2 \cdots \int_{x_{k-1}}^{z_{k-1}} a(x_k)(1 - A(x_k))^{n-k}dx_k,
$$

that ends the proof. $\square$

As a consequence of the theorem, the main system reliability characteristics can be calculated.

Remark 1. *Based on the distribution of the system's time to failure, any other system's reliability characteristics can be calculated, such as:*

- *its reliability function $R(y) = 1 - F_Y(y)$;*
- *its mean lifetime $\mathbf{E}[T] = \int_0^\infty R(t)dt$;*
- *its lifetime variation $\mathbf{var}[T]$.*

3.4. A Special Case: Exponential Distribution

In a special case, when the system components' A_i $(i = \overline{1,n})$ lifetimes have exponential (Exp) distribution with a parameter α the integral (6) can be calculated analytically, but the calculations are rather cumbersome. We show it for the given value of $k = 2$. But for exponential distribution of the system components' lifetime, we propose another approach for the system lifetime distribution. It is based on the memoryless property of any exponentially distributed rv.

Denote by T_i the time interval between $i-1$-th and i-th components failures, $i = \overline{1, k-1}$ $(T_0 = 0)$. Then due to the memoryless property of the exponential distribution the time to the k-th failure Y_k is the sum

$$
Y_k = T_1 + T_2 + \cdots + T_k,
$$

of k independent exponentially distributed rv T_i with parameters

$$
\lambda_1 = n\alpha, \quad \lambda_i = c_1c_2 \cdots c_{i-1}(n-i+1)\alpha = \bar{c}_i(n-i+1)\alpha, \ i = \overline{2,k},
$$

where for simplicity additional notations are used:

$$
\bar{c}_i = \begin{cases} 1, & i = 1, \\ c_1 \cdots c_{i-1}, & i = \overline{2,k}. \end{cases}
$$

The moment generating function (mgf) of the system's lifetime in this case has the following form:

$$\phi_k(s) = \mathbf{E}\left[e^{-sY_k}\right] = \prod_{1 \leq i \leq k} \mathbf{E}\left[e^{-sT_i}\right] = \prod_{1 \leq i \leq k} \frac{\bar{c}_i \lambda_i}{s + \bar{c}_i \lambda_i}.$$

To apply the above theorem and the proposed approach, let us consider the simplest example of a k-out-of-n model with $k = 2$. In this case, suppose $c_1 = c$. Thus, according to (1) the joint distribution of rv $X_{(1)}$, $X_{(2)}$ is

$$f_2(x_1, x_2) = \frac{n!}{(n-2)!} a(x_1) a(x_2)(1 - A(x_2))^{n-2} = \frac{n!}{(n-2)!} \alpha^2 e^{-\alpha x_1} e^{-(n-1)\alpha x_2}.$$

Calculate cdf $F_{Y_2}(y)$ of the rv $Y_2 = (1-c)X_{(1)} + cX_{(2)}$,

$$
\begin{aligned}
F_{Y_2}(y) &= \mathbf{P}\{(1-c)X_{(1)} + cX_{(2)} < y\} = \mathbf{P}\left\{X_{(2)} < \frac{y - (1-c)X_{(1)}}{c}\right\} \\
&= n(n-1)\alpha^2 \int_0^y e^{-\alpha x_1} dx_1 \int_{x_1}^{\frac{y-(1-c)x_1}{c}} e^{-(n-1)\alpha x_2} dx_2 \\
&= 1 + \frac{n-1}{nc - (n-1)} e^{-n\alpha y} - \frac{nc}{nc - (n-1)} e^{-\frac{(n-1)\alpha}{c} y},
\end{aligned}
$$

and therefore its pdf for $y \geq 0$ is

$$f_{Y_2}(y) = \frac{n(n-1)\alpha}{nc - (n-1)} \left(e^{-\frac{(n-1)\alpha}{c} y} - e^{-n\alpha y}\right).$$

Please note that this result holds for $c \neq (n-1)/n$ and in this case the distribution is a mixture of exponential distributions. The point $c = (n-1)/n$ is a singular point for which cdf of the rv Y_2 is the Erlang distribution,

$$F_{Y_2}(y) = 1 - e^{-n\alpha y} - n\alpha y e^{-n\alpha y}, \quad y > 0,$$

with pdf

$$p_{Y_2}(y) = n^2 \lambda^2 y e^{-n\lambda y}, \quad y > 0.$$

Remark 2. *The singularity in the calculation of the cdf of the system's lifetime arises because for some special values of the coefficient c_i (here for $c = (n-1)/n$) the moment generating function of the system's lifetime has multiple roots that leads to changing of the shape of distribution.*

With the help of another approach one can find mgf of the system's lifetime in the following form:

$$\phi_2(s) = \frac{n(n-1)\alpha^2}{s^2 + (2n-1)\alpha s + n(n-1)\alpha^2}.$$

By expanding this expression into simple fractions, we find

$$\phi_2(s) = \frac{n(n-1)\alpha}{s + n\alpha} - \frac{n(n-1)\alpha}{s + (n-1)\alpha},$$

then, by calculating the inverse function, we obtain

$$f_2(y) = n(n-1)\alpha\left(e^{-(n-1)\alpha y} - e^{-n\alpha y}\right),$$

which is the same as the result above for $c = 1$.

The analytical calculations of the reliability characteristics are not always possible. Nevertheless, their numerical analysis in the wide domain of initial data is possible. Therefore, in the next section a procedure for the numerical calculation of different reliability characteristics of the considered system will be proposed. Furthermore, in Section 5 this procedure will be used for the numerical analysis of the model with some examples.

4. The General Calculation Procedure of the System Reliability Characteristics and Numerical Experiments

Based on the results of the previous section, the general procedure for the problem solution can be implemented with the help of the following algorithm (Algorithm 1).

Algorithm 1 : General algorithm for calculation of reliability function

---

Beginning. Determine: Integers n, k, real c_i $(i = \overline{1,k})$, distribution $A(t)$ of the system components' lifetime and its pdf.

Step 1. Taking into account that the system's failure moment Y_k according to formula (2) equals

$$Y_k = C_1 X_1 + C_2 X_2 + \cdots + C_{k-1} X_{k-1} + C_k X_k,$$

calculate the following,

$$C_i = \begin{cases} 1 - c_i, & i = 1, \\ c_1 \cdots c_{i-1}(1 - c_i) & i = \overline{2, k-1}, \\ c_1 \cdots c_{k-1} & i = k. \end{cases}$$

Step 2. Taking into account that according to formula (1), the joint distribution density of first k order statistics $X_1 \le X_2 \le \cdots \le X_k$ holds

$$f_{X_1 X_2 \ldots X_k}(x_1, x_2, \ldots, x_k) = \frac{n!}{(n-k)!} a(x_1) a(x_2) \ldots a(x_k)(1 - A(x_k))^{n-k},$$

with which following to (6) calculate the reliability function

$$R(y) = 1 - F_{Y_k}(y) = 1 - \frac{n!}{(n-k)!} \int_0^y a(x_1) dx_1 \int_{x_1}^{z_1} a(x_2) dx_2 \cdots \int_{x_{k-1}}^{z_{k-1}} a(x_k)(1 - A(x_k))^{n-k} dx_k,$$

where the limits of integration are determined by the relation (5)

$$z_0 = y, \quad z_i = z_i(y; x_1, \ldots, x_i) = \frac{y - C_1 x_1 + C_2 x_2 - \ldots - C_i x_i}{c_1 c_2 \ldots c_i} \quad (i = \overline{1, k-1}).$$

Find the values of the constants c_i (singular points at which the denominator of the cdf $F_{Y_k}(y)$ turns into 0) for which the cdf changes its appearance.

Step 3. From the system reliability function $R(y)$, calculate
– mean time to the system failure

$$\mu_T = \mathbf{E}[Y_k] = \int_0^\infty R(y) dy;$$

– its variance

$$\sigma_T^2 = \mathbf{Var}[Y_k] = \int_0^\infty (y - \mu_T)^2 f(y) dy, \quad where \quad f(y) = \frac{d}{dy} F_{Y_k}(y),$$

and coefficient of variation

$$v = \frac{\sigma}{\mu}.$$

Stop.

Remark 3. *The algorithm can also be used to solve other different problems, for example, to analyze the sensitivity of the system's reliability function and its characteristics to the shape of the lifetime distribution of the system's components.*

Furthermore, the algorithm will be applied to some examples.

5. Numerical Experiments: 2-Out-Of-6 System

According to Algorithm 1, we calculate the reliability function of a 2-out-of-6 system. Since such a system fails due to the failure of two components, we have only one constant that defines the decreasing residual lifetime of surviving components. Therefore, hereafter, we suppose $c_1 = c$. Consider the Gnedenko–Weibull (*GW*) distribution as the lifetime distribution of the system's components, $A(t) \sim GW\left(\theta, \frac{a}{\Gamma(1+\theta^{-1})}\right)$, with the corresponding cdf

$$A(t) = 1 - \exp\left\{-\left(\frac{t\Gamma(1+\theta^{-1})}{a}\right)^{\theta}\right\}, t > 0,$$

where

- a is a fixed mean components' lifetime,
- θ is the shape parameter of *GW* distribution calculated based on the preset value of the coefficient of variation,
- $v = \dfrac{\sigma}{a} = a^{-1} \cdot \sqrt{\dfrac{\Gamma(1+2\cdot\theta^{-1})}{\Gamma(1+\theta^{-1})^2} - 1}$ is the coefficient of variation,
- σ is the standard deviation.

Additionally, consider the Erlang (*Erl*) distribution, $A(t) \sim Erl(l,\theta)$ with pdf

$$a(y) = \frac{\theta^l}{\Gamma(l)}y^{l-1}e^{-\theta y}, y > 0.$$

In this case, the distribution's parameters can be represented via the corresponding mean a and coefficient of variation v as follows,

$$l = v^{-2}, \quad \theta = (av^2)^{-1}.$$

For numerical experiments, we consider the reliability function and its characteristics of a 2-out-of-6 system for given distributions with a fixed mean a and different values of v. Thus, we can analyze the influence of the coefficient of variation of the repair time on the reliability characteristics of the system. In other words, investigate its sensitivity.

Suppose that the mean lifetime of the component $a = 1$. If $\theta = 1$, *GW* and *Erl* distributions transform into the exponential one with the mean time a and the coefficient of variation $v = 1$. In this case, its reliability function is

$$R(t) = \frac{5e^{-6t} - 6c \cdot e^{-\frac{5t}{c}}}{5 - 6c}. \tag{8}$$

From Formula (8) it is clear that $c = \frac{5}{6}$ leads to changing of the shape of distribution.

Since calculating the coefficient θ for *GW* through the value of v is quite difficult, we define the parameter θ so that $v \approx 0.5$. Moreover, if θ of *GW* takes non-integer values, it is not always possible to obtain a closed-form reliability function $R(t)$ according to Algorithm 1 (the integrand takes a complex form). Therefore, define $\theta = 2$, then, the coefficient of variation $v = 0.5227$. For *Erl* distribution, suppose that $v = 0.5$, which leads to $\theta = 4$.

Suppose $c = 0.1; 0.5; 1$. Figure 3 illustrates the reliability function of the 2-out-of-6 system for different distributions, as well as c and v. Here, solid line means $v = 1$ and reliability function (8), dashed one is for *GW* with $v = 0.5227$ and dash-dotted is for *Erl* with $v = 0.5$. The legend of the figure denotes the color of line for different c.

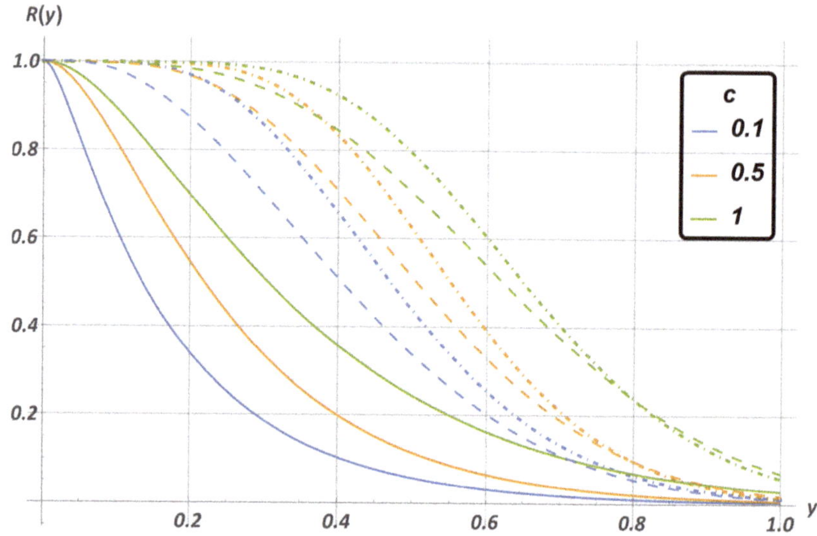

Figure 3. Reliability function of a 2-out-of-6 system.

The figure shows that the higher reliability coincides with the lower value of v. The case $c = 1$ means the absence of load from the failed components to the surviving ones, thus this case corresponds to the highest reliability for different v compared to the values $c < 1$. Moreover, dependence of the reliability function curve on the shape of lifetime distribution is observed. On a small interval y, the system reliability as $A \sim Erl$ is higher than as $A \sim GW$ for each c, despite close values v. This may indicate the sensitivity of the reliability function not only to the shape of the lifetime distribution, but also to the corresponding value of the coefficient of variation.

According to the algorithm, we calculate other reliability characteristics of the 2-out-of-6 system (Tables 1 and 2). These characteristics correspond to the system's reliability behavior, shown in Figure 3. The lower value of v leads to the higher value of the system lifetime expectation $\mathbf{E}[Y_2]$, and the lower value of c leads to the lower value of $\mathbf{E}[Y_2]$. Moreover, as $v \approx 0.5$ the relative error between the considered distributions is 14.11% for $c = 0.1$, 7.98% for $c = 0.5$ and 3.86% for $c = 1$.

Table 1. $\mathbf{E}[Y_2]$ of a 2-out-of-6 system.

	$c = 0.1$	$c = 0.5$	$c = 1$
$v = 0.5\ (A \sim Erl)$	0.4925	0.5670	0.6668
$v = 0.5227\ (A \sim GW)$	0.4316	0.5251	0.6420
$v = 1\ (A \sim Exp)$	0.1867	0.2667	0.3667

To distinguish coefficients of variation of the components and the whole system, denote them as v_{comp} and v_{sys}, respectively. Thus, Table 2 shows the following. With a decrease in c, the coefficient of variation of the system v_{sys} grows and tends to the value of the coefficient of variation of each system component v_{comp}. The increasing v_{comp} leads to the increasing v_{sys} for all distributions and c. Thus, the coefficient of variation of the system v_{sys} confirms that as c tends to 0 and v_{comp} tends to 1, variability with respect to the average lifetime of the system $\mathbf{E}[Y_2]$ grows.

Table 2. v_{sys} of a 2-out-of-6 system.

	$c = 0.1$	$c = 0.5$	$c = 1$
$v_{comp} = 0.5\ (A \sim Erl)$	0.3802	0.3049	0.2986
$v_{comp} = 0.5227\ (A \sim GW)$	0.4813	0.3854	0.3641
$v_{comp} = 1\ (A \sim Exp)$	0.8993	0.7289	0.7100

Table 1 showed that with increasing c and $v \approx 0.5$, the mean system lifetime $\mathbf{E}[Y_2]$ is very close. However, Figure 3 shows that over a small interval y with these c and v, the reliability of the system has significant differences. This leads to the study the quantiles of the system reliability. This measure shows how long the system will be reliable with a fixed probability. The quantiles $q_\gamma = R^{-1}(\gamma)$ of the reliability function are presented in Figures 4–6. In all cases, red bullets correspond to $\gamma = 0.99$, whereas black bullets correspond to $\gamma = 0.9$.

All the values for quantiles $\gamma = 0.999; 0.99; 0.9$ are presented in Table 3 for different distributions. The values in the table show that for the presented quantiles q_γ, the shape of the lifetime distribution of the system's components as well as its coefficient of variation play a critical role on the system's reliability. Therefore, for example, as $c = 0.1$ and $A \sim Erl$ a given reliability level 0.9 will last about 8 times longer than for $c = 0.1$ and $A \sim Exp$. At that for $q_{0.999}$, the difference for similar case is almost 40 times. As the coefficient c increases, this difference decreases for all values of the quantiles and lifetime distributions of the components. As $c = 1$ this difference is reduced by about two times. Thus, even as $c = 1$, which defines no changing in components' residual lifetimes, the influence not only of the lifetime distribution of the components but also its coefficient of variation on the reliability of the system is huge. This once again confirms the sensitivity of the reliability characteristics of the k-out-of-n system to the shape of the lifetime distribution and the coefficient of variation of system's components.

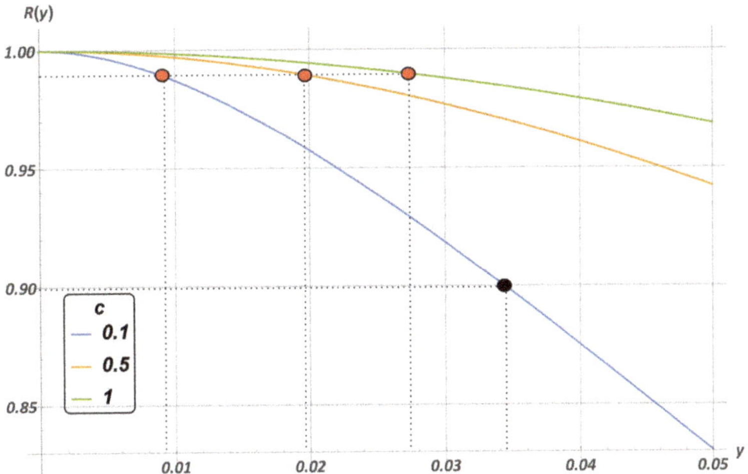

Figure 4. Reliability function with $v = 1$ and quantiles ($A \sim Exp$). Red and black bullets are the points of intersection of the reliability function curves with fixed reliability levels of 0.99 and 0.9, respectively.

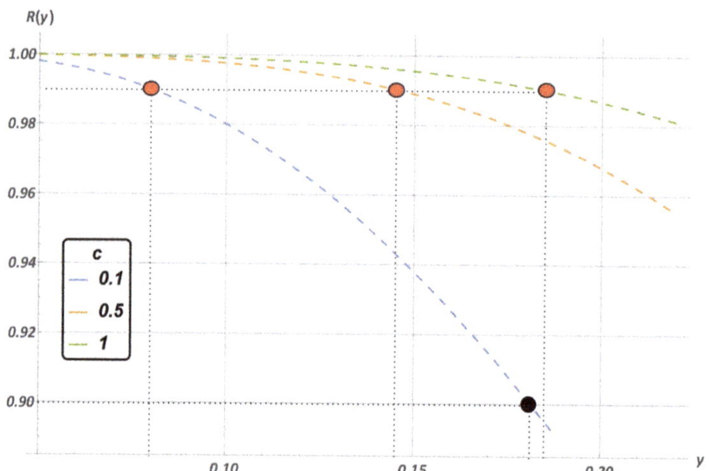

Figure 5. Reliability function with $v = 0.5227$ and quantiles ($A \sim GW$). Red and black bullets are the points of intersection of the reliability function curves with fixed reliability levels of 0.99 and 0.9, respectively.

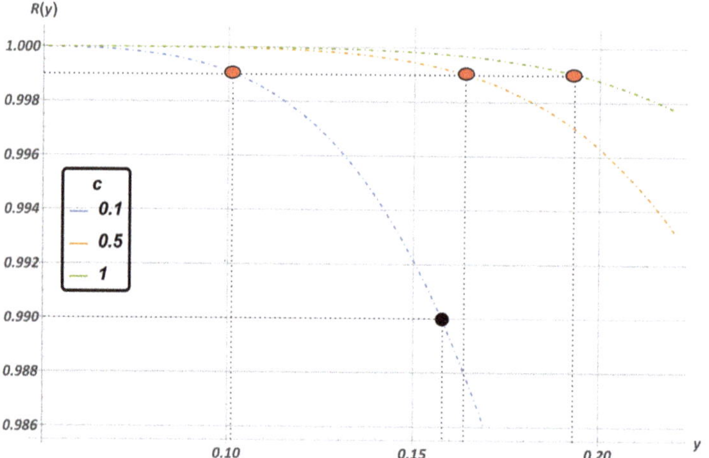

Figure 6. Reliability function with $v = 0.5$ and quantiles ($A \sim Erl$). Red and black bullets are the points of intersection of the reliability function curves with fixed reliability levels of 0.99 and 0.9, respectively.

Table 3. Quantiles of reliability function q_γ.

		$c = 0.1$	$c = 0.5$	$c = 1$
$q_{0.999}$	$A \sim Exp$	0.0026	0.0059	0.0083
	$A \sim GW$	0.0419	0.0804	0.1027
	$A \sim Erl$	0.1019	0.1641	0.1945
$q_{0.99}$	$A \sim Exp$	0.0088	0.0192	0.0271
	$A \sim GW$	0.0804	0.1458	0.1859
	$A \sim Erl$	0.1576	0.2345	0.2784
$q_{0.9}$	$A \sim Exp$	0.0344	0.0691	0.0972
	$A \sim GW$	0.1813	0.2784	0.3517
	$A \sim Erl$	0.271	0.3574	0.424

6. Conclusions and the Further Investigations

The reliability function of a new k-out-of-$n : F$ model is investigated, under the new assumptions that the failures of its components lead to the increase in the load on the remaining ones and, consequently, to the change in their residual lifetimes. To model the situation, we proposed a novel approach based on the transformation of the order statistics of the system components' lifetimes , which is the new field of application of order statistics. An algorithm for calculation of the system's reliability function and its moments has been developed. Numerical experiments for the special case of the considered model based on the real-world systems have been carried out. The experiments show an essential sensitivity of the model reliability function and its moments to the shapes of the lifetime distributions of the system's components and their coefficient of variation.

Furthermore, it is proposed we extend this approach to the investigation of stationary characteristics of the model and consider its preventive maintenance, aiming to improve its reliability characteristics.

Author Contributions: Conceptualization, writing–original draft preparation, supervision, project administration, V.R.; validation, investigation, visualization, N.I.; writing–review and editing, methodology, D.K.; data curation, software, T.M. All authors have read and agreed to the published version of the manuscript.

Funding: This paper has been supported by the RUDN University Strategic Academic Leadership Program (recipients V.R, supervision and problem setting, N.I., visualization, D.K. writing–review and editing, T.M., analytic results). This paper has been partially funded by RFBR according to the research projects No.20-01-00575A (recipients V.R., conceptualization, and N.I., formal analysis) and RSF according to the research projects No.22-49-02023 (recipient N.I., validation, D.K. review and analytic results).

Institutional Review Board Statement: Not applicable.

Informed Consent Statement: Not applicable.

Data Availability Statement: Not applicable.

Acknowledgments: The authors express their gratitude to the Referees for the valuable suggestions, which improved the quality of the paper.

Conflicts of Interest: The authors declare no conflict of interest.

Abbreviations

The following abbreviations are used in this manuscript:

iid	independent and identically distributed
rv	random variable
cdf	cumulative distribution function
pdf	probability density function
UAV	Unmanned Aerial Vehicle
UUV	Unmanned Underwater Vehicle
mgf	moment generating function
Exp	Exponential distribution
GW	Gnedenko–Weibull distribution
Erl	Erlang distribution

References

1. Trivedi, K.S. *Probability and Statistics with Reliability, Queuing and Computer Science Applications*, 2nd ed.; John Wiley & Sons: New York, NY, USA, 2016. [CrossRef]
2. Chakravarthy, S.R.; Krishnamoorthy, A.; Ushakumari, P.V. A k-out-of-n reliability system with an unreliable server and Phase type repairs and services: The (N, T) policy. *J. Appl. Math. Stoch. Anal.* **2001**, *14*, 361–380. [CrossRef]
3. Rykov, V.; Kozyrev, D.; Filimonov, A.; Ivanova, N. On Reliability Function of a k-out-of-n System with General Repair Time Distribution. *Probab. Eng. Inf. Sci.* **2020**, *35*, 885–902. [CrossRef]

4. Pascual-Ortigosa, P.; Sáenz-de-Cabezón, E. Algebraic Analysis of Variants of Multi-State k-out-of-n Systems. *Mathematics* **2021**, *9*, 2042. [CrossRef]
5. Zhang, T.; Xie, M.; Horigome, M. Availability and reliability of (k-out-of-($M + N$)): Warm standby systems. *Reliab. Eng. Syst. Saf.* **2006**, *91*, 381–387. [CrossRef]
6. Gertsbakh, I.; Shpungin, Y. Reliability Of Heterogeneous ((k, r)-out-of-(n, m)) System. *RTA* **2016**, *3*, 8–10.
7. Ushakov, I. A universal generating function. *Sov. J. Comput. Syst. Sci.* **1986**, *24*, 37–49.
8. Ushakov, I. Optimal standby problem and a universal generating function. *Sov. J. Comput. Syst. Sci.* **1987**, *25*, 61–73.
9. Levitin, G. *The Universal Generating Function in Reliability Analysis and Optimization*; Springer Series in Reliability Engineering; Springer: London, UK, 2005. [CrossRef]
10. Kala, Z. New Importance Measures Based on Failure Probability in Global Sensitivity Analysis of Reliability. *Mathematics* **2021**, *9*, 2425. [CrossRef]
11. Rykov, V.; Sukharev, M.; Itkin, V. Investigations of the Potential Application of k-out-of-n Systems in Oil and Gas Industry Objects. *J. Mar. Sci. Eng.* **2020**, *8*, 928. [CrossRef]
12. Rykov, V.; Kochueva, O.; Farkhadov, M. Preventive Maintenance of a k-out-of-n System with Applications in Subsea Pipeline Monitoring. *J. Mar. Sci. Eng.* **2021**, *9*, 85. [CrossRef]
13. Vishnevsky, V.M.; Kozyrev, D.V.; Rykov, V.V.; Nguyen, D.P. Reliability modeling of an unmanned high-altitude module of a tethered telecommunication platform. *Inf. Technol. Comput. Syst.* **2020**, *4*, 26–36. [CrossRef]
14. Zhang, J.; Jiang, Y.; Li, X.; Huo, M.; Luo, H.; Yin, S. An adaptive remaining useful life prediction approach for single battery with unlabeled small sample data and parameter uncertainty, *Reliab. Eng. Syst. Saf.* **2022**, *222*, 108357. [CrossRef]
15. Zhang, J.; Jiang, Y.; Li, X.; Luo, H.; Yin, S.; Kaynak, O. Remaining Useful Life Prediction of Lithium-Ion Battery with Adaptive Noise Estimation and Capacity Regeneration Detection. *IEEE/ASME Trans. Mechatron.* **2022**, 1–12 [CrossRef]
16. Zhang, J.; Jiang, Y.; Wu, S.; Li, X.; Luo, H.; Yin, S. Prediction of remaining useful life based on bidirectional gated recurrent unit with temporal self-attention mechanism, *Reliab. Eng. Syst. Saf.* **2022**, *221*, 108297. [CrossRef]
17. Eryilmaz, S. Phase type stress-strength models with reliability applications. *Commun. Stat.—Simul. Comput.* **2018**, *47*, 954–963. [CrossRef]
18. Bai, X.; Shi, Y.; Liu, Y.; Liu, B. Reliability estimation of stress-strength model using finite mixture distributions under progressively interval censoring. *J. Comput. Appl. Math.* **2019**, *348*, 509–524. [CrossRef]
19. Zhang, L.; Xu, A.; An, L.; Li, M. Bayesian inference of system reliability for multicomponent stress-strength model under Marshall-Olkin Weibull distribution. *Systems* **2022**, *10*, 196. [CrossRef]
20. Tang, Y.; Zhang, J. New model for load-sharing k-out-of-n : G system with different components. *J. Syst. Eng. Electron.* **2008**, *19*, 842, 748—751. [CrossRef]
21. Hellmich, M. Semi-Markov embeddable reliability structures and applications to load-sharing k-out-of-n system. *Int. J. Reliab. Qual. Saf. Eng.* **2013**, *20*, 1350007. [CrossRef]
22. Bairamov, I.; Arnold, B.C. On the residual lifelengths of the remaining components in an $n - k + 1$ out of n system. *Stat. Probab. Lett.* **2008**, *78*, 945–952. [CrossRef]
23. Nguyen, D.P.; Kozyrev, D.V. Reliability Analysis of a Multirotor Flight Module of a High-altitude Telecommunications Platform Operating in a Random Environment. In Proceedings of the 2020 International Conference Engineering and Telecommunication (En&T), Dolgoprudny, Russia, 25–26 November 2020, pp. 1–5. [CrossRef]
24. Rykov, V.; Ivanova, N.; Kochetkova, I. Reliability Analysis of a Load-Sharing k-out-of-n System Due to Its Components' Failure. *Mathematics* **2022**, *10*, 2457. [CrossRef]
25. Katzur, A.; Kamps, U. Order statistics with memory: A model with reliability applications. *J. Appl. Probab.* **2016**, *53*, 974–988. [CrossRef]
26. Cramer, E.; Kamps, U. Sequential order statistics and k-out-of-n systems with sequentially adjusted failure rates. *Ann. Inst. Stat. Math.* **1996**, *48*, 535–549. [CrossRef]
27. Navarro, J.; Marco, B. Coherent Systems Based on Sequential Order Statistics. *Nav. Res. Logist.* **2011**, *58*, 123–135. [CrossRef]
28. Sutar, S.; Naik-Nimbalkar, U.V. A load share model for non-identical components of a k-out-of-m system. *Appl. Math. Model.* **2019**, *72*, 486–498. [CrossRef]
29. Kozyrev, D.V.; Phuong, N.D.; Houankpo, N.G.K.; Sokolov, A. Reliability evaluation of a hexacopter-based flight module of a tethered unmanned high-altitude platform, *Commun. Comput. Inf. Sci.* **2019**, *1141*, 646–656._52. [CrossRef]
30. David, H. A.; Nagaraja, H. N. *Order Statistics*, 3rd ed.; John Wiley & Sons: New York, NY, USA, 2003. [CrossRef]

MDPI

St. Alban-Anlage 66

4052 Basel

Switzerland

Tel. +41 61 683 77 34

Fax +41 61 302 89 18

www.mdpi.com

Mathematics Editorial Office

E-mail: mathematics@mdpi.com

www.mdpi.com/journal/mathematics